T0135499

Frank H. Scharf

Fluid Dynamic and Kinetic Modeling of the Near-Cathode Region in Thermal Plasmas

Fluid Dynamic and Kinetic Modeling of the Near-Cathode Region in Thermal Plasmas

Frank H. Scharf

Dissertation zur Erlangung des Grades eines Doktor-Ingenieurs der Fakultät für Elektrotechnik und Informationstechnik an der Ruhr-Universität Bochum

Bochum, 30. September 2008

Ruhr–Universität Bochum, Lehrstuhl für Theoretische Elektrotechnik

Prof. Dr. rer. nat. Ralf Peter Brinkmann

Universitätsstraße 150, 44801 Bochum, Deutschland

Bibliografische Information der Deutschen Nationalbibliothek

Die Deutsche Nationalbibliothek verzeichnet diese Publikation in der
Deutschen Nationalbibliografie; detaillierte bibliografische Daten sind
im Internet über http://dnb.d-nb.de abrufbar.

ISBN 978-3-8325-2182-0

Logos Verlag Berlin GmbH
Comeniushof, Gubener Str. 47,
10243 Berlin
Tel.: +49 030 42 85 10 90
Fax: +49 030 42 85 10 92
INTERNET: http://www.logos-verlag.de

Contents

List of Figures

Nomenclature

All units are given for the non-normalized variables. Vectors are represented by bold letters, tensors and matrices are written in bold letters with two underlines.

Latin symbols:

$a_{\#,\#(,\#)}$	sum coefficient (dimensionless)
$b_{\#,\#(,\#)}$	sum coefficient (dimensionless)
$\boldsymbol{c}$	velocity vector (m s^{-1})
e	electron charge ($-1.602 \cdot 10^{-19}$ C), also Euler's number (2.71828)
$e_{\#}$	kinetic energy with subscript ions/atoms (kg m^2 s^{-2})
$\boldsymbol{E}, E$	electric field (V m^{-1})
E_i	ionization energy (eV)
$f_{\#}$	particle distribution function with subscript to denote species (s^3 m^{-6})
h	cell width used in discretization (dimensionless)
k	cell index used in discretization (dimensionless)
k_i	ionization rate (m^4 s^{-2})
k_r	recombination rate (m^7 s^{-2})
K	upper limit for indices k and κ (dimensionless)
L	typical system size (m), also discretization interval size (dimensionless)
$m, \widetilde{m}, \widetilde{\widetilde{m}}$	sum indices (dimensionless)
$m_{\#}$	particle mass with subscript ions/atoms/electrons (kg)
M	upper limit for $m, \widetilde{m}, \widetilde{\widetilde{m}}$ (dimensionless)
$\underline{\underline{M}}$	linearization matrix (dimensionless)
$n, \widetilde{n}, \widetilde{\widetilde{n}}$	sum indices (dimensionless)
$n_{\#}$	density with subscript ions/atoms/electrons (m^{-3})
$n_{\#p}$	corresponding density in the Saha plasma (m^{-3})
N	upper limit for $n, \widetilde{n}, \widetilde{\widetilde{n}}$ (dimensionless)
$\underline{\underline{N}}$	linearization matrix (dimensionless)
$p, \underline{\boldsymbol{P}}$	total pressure (kg m^{-1} s^{-2})
q_s	particle charge of species s (C)
q	heat flux density (kg s^{-3})
R	universal gas constant (8.314472 J K^{-1} mol^{-1})

Nomenclature

$T_\#$	temperature, subscript $\underline{c}$athode/$\underline{i}$ons/$\underline{a}$toms/$\underline{e}$lectrons/$\underline{h}$eavy particles (eV)
$T_{a,0}$	neutral temperature at sheat edge (eV)
$\underline{\underline{T}}$	linearization matrix (dimensionless)
s	curve parameter (dimensionless)
$\boldsymbol{u}_\#$	average velocity with subscript $\underline{i}$ons/$\underline{a}$toms (m s^{-1})
$v_\#$	velocity with subscript $\underline{t}$hermal/$\underline{i}$ons/$\underline{a}$toms/$\underline{e}$lectrons (m s^{-1})
$v_{\#p}$	corresponding velocity in the Saha plasma (m s^{-1})
v_a^*	fixed value of v_a, used in linearization (dimensionless)
v_{as}	sound velocity for neutral atoms (m s^{-1})
v_B	Bohm velocity (m s^{-1})
v_i^*	fixed value of v_i, used in linearization (dimensionless)
$\boldsymbol{v}_{\lambda 1}, \boldsymbol{v}_{\lambda 2}$	eigenvectors (dimensionless)
$W_{\#,\#}$	particle collision probability; species given by subscripts (dimensionless)
z	Cartesian coordinate (m)
$Z_\#$	partition sum with subscript $\underline{i}$ons/$\underline{a}$toms

Greek Symbols:

α	fluid model normalization constant, details in section 2.2 (dimensionless)
α_{cr}	special, critical value of α (dimensionless)
$\alpha_{\#\#}$	kinetic equivalent of α (dimensionless)
β	normalization constant, equal to T_e/T_h (dimensionless)
γ	normalization constant, equal to $n_{ip}k_r/k_i$ (dimensionless)
Γ	flux vector (kg s^{-1} m^{-2} for mass flux, kg s^{-3} for energy flux)
$\delta_{m,n}$	Kronecker Delta; equal to unity for $m = n$, zero otherwise (dimensionless)
ζ	normalized Cartesian coordinate (dimensionless)
ϑ	temperature ratio, details in section 5.2.2 (dimensionless)
θ	spherical coordinate (rad)
κ	cell index used in discretization (dimensionless)
λ_D	Debye length (m)
λ_i	ionization length (m)
$\lambda_{\#\#}$	mean free path for collisions between two particles (m)
λ_s	eigenvalues, $s \in \mathbb{N}$ (dimensionless)
μ	mass ratio (dimensionless)
ν_1	frequency of collisions between electrons and heavy particles (s^{-1})
$\bar{\nu}_1$	normalized version of ν_1 (dimensionless)
ν_{ia}	frequency of collisions between ions and neutral atoms (s^{-1})
$\underline{\underline{\Pi}}$	momentum flux density (kg m^{-1} s^{-2})
φ	spherical coordinate (rad)
Φ	electric potential (V)
Ω	solid angle unit (sr)

1. Introduction

1.1. Modeling of High Intensity Discharges

Plasma modeling and simulations have become an important and widely-used tool in both research and industrial product development. Although simulations never can completely replace experiments, they are often more time and cost effective. In addition, they sometimes allow insights into physical properties that are (yet) inaccessible to measurements. One branch of industry employing plasma simulations is the lighting industry. We encounter its products every day – and according to a 2007 study by the German Ministry of Education and Research [www1], the world market value in 2005 was 18.5 billion Euros with an expected long-term growth rate of 5.5 percent (versus US$ 11 billion in 2002 according to [1], chapter 34). These 18.5 billion Euros have to be distributed among various technology branches, like traditional incandescent lamps, compact fluorescent lamps (CFLs), lasers, (organic) light emitting diodes ((O)LEDs), and high intensity discharge (HID) lamps.

HID lamps – the piece of interest for this work – only represent about 1 percent of the total market value (still approximately 200 million Euros), but almost 50 percent of the total lumen production [2]. HID lamps can be found in automotive head lights, video projectors, or in large area lighting applications such as production sites or stadiums. Despite the rise of LEDs, HID lamps remain competitive especially in these areas. Additionally, approaches to extend the application range of HID lamps (e.g., for home lighting purposes) are currently investigated.

Figure 1.1 shows a photograph of a commercial HID lamp and figure 1.2 shows a schematic diagram of the inner discharge vessel. In the figures, two electrodes reach into the vessel, which is usually made of quartz glass or ceramics. The vessel is filled with a gas (xenon, for example) at typically 10 bar, mercury, and some additives. The mercury, among other functions, leads to a low ignition voltage and a high operation voltage, resulting in an improved efficiency. Since mercury is hazardous, significant effort is currently put into finding a replacement. The additives usually consist of metals (often zinc or sodium), halides (often iodine), rare earths (often dysprosium), and combinations thereof. Their purpose is to increase the lamp efficiency as well, but also to control the color rendering abilities of the lamp. Upon ignition, the mercury and the

1

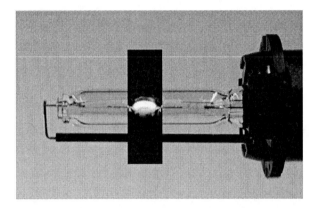

Figure 1.1.: Photograph of a commercial HID lamp (so-called D-lamp for automotive purposes). The discharge vessel is the small glass capsule inside the darkened area. Courtesy of AEPT, Ruhr-Universität Bochum.

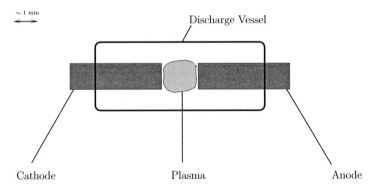

~ 1 mm

Discharge Vessel

Cathode   Plasma   Anode

Figure 1.2.: Schematic setup of an HID lamp. Two electrodes reach into the discharge vessel. The voltage applied to the electrodes sustains a light-emitting plasma in between. The setup is usually enclosed in a second vessel for protection and improved efficiency.

additives evaporate, raising the pressure to typically 20 − 100 bar. A light emitting plasma starts burning between the electrodes. The way the plasma is attached to the electrodes is rather complex. In general, two fundamentally different attachment forms can be observed at the cathode, usually referred to as diffuse and spot mode (cf. figure 1.3). The mechanisms for the formation of either mode are of great interest from both an academic and an industrial standing point but not completely understood yet. Modeling and simulation is one method of developing better understanding.

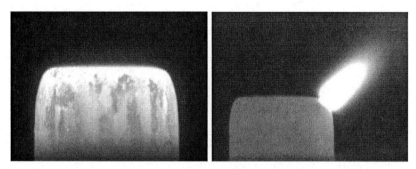

Figure 1.3.: Photographs of the two observed modes of arc attachment: diffuse (left) and spot (right). Courtesy of AEPT, Ruhr-Universität Bochum.

To model the near-cathode region it is useful to introduce a layer (also called "sheath") in front of the cathode to describe the transition from the cathode to the plasma column. This layer is often further subdivided into at least a space charge sheath and a presheath. A detailed discussion of this is given at the beginning of chapter 2 and in [3].

It is possible, in principle, to operate such a system with direct current (DC). However, cataphoretic effects cause a different composition of the plasma in front of the cathode and anode, leading to different light emission properties. Moreover, depending upon the current amplitude, the erosion of the cathode and anode may be different, limiting the lifetime of the electrodes. To remove these shortcomings, HID lamps are most commonly driven by some form of alternating current (AC). The current shape and frequency vary. One often finds a low-frequency (50 Hz) switched DC, but more complex wave forms and higher frequencies (up to a few 100 kHz) are also applied and investigated for improved lamp characteristics. Especially for the low-frequency cases, a stationary model is usually sufficient.

Further assumptions and (necessary) simplifications should be discussed on a case-by-case basis, since they depend strongly upon additional factors, such as the complexity and purpose of the respective lamp model. At this point, only the general kinds of possible models shall be discussed. One possibility is a straightforward approach that investigates the whole lamp as one system. Models following such an approach exist (for example, [4]-[7]). However, in order to be handeable, such models often neglect a many details. The alternative is to split the whole system up into several interfacing modules, which is not always possible and almost never a trivial task. If, perchance, the separation succeeds, the modules can be treated separately in more detail and later recombined. In the case of HID lamps, such a separation has been shown to be possible and corresponding models exist, see for example [3] and [8] - [14]. These models separate the whole system into several modules: one for the cathode, another one for the near-cathode layer, and some others for the rest of the lamp. For the treatment of the cathode module, simply solving a heat conduction problem is sufficient. The module describing the near-cathode boundary layer is much more complex. It is this module and how to model it that this thesis focuses on. The interface between the two modules are so-called transfer functions, namely the power flux density q and the current density j. They can be calculated from the model of the near-cathode layer and then used as boundary conditions for the heat conduction problem, thus linking both models.

The aim of this thesis is to develop a model of the near-cathode layer that advances the work done so far by others, so that a consistent description is possible. Suitable tools for this task are fluid dynamics or kinetic theory which are briefly described in sections 1.2 and 1.3. Particle in cell (PIC) simulations are not appropriate because of the high pressures usually present in HID lamps (up to 200 bar). Chapters 2 and 3 show a detailed study of fluid dynamic modeling of the cathode sheath. Based on the results from those chapters, it was decided to apply kinetic methods instead. The corresponding kinetic model and some exemplary results are presented in chapter 4 and section 5.4.

It should be noted that the sheath in front of the anode has completely different properties from the cathode sheath, results from cathode models cannot be transferred to the anode sheath (cf. [15] and references therein, in particular [16]). However, cathode models are not only useful for lamps, but also for e.g. magnetohydrodynamic (MHD) generators, so that the results are useful for more than a single application.

1.2. Fluid Dynamic Modeling

Webster's New Encyclopedic Dictionary [17] defines the noun *fluid* as "a substance tend-ing to flow or conform to the outline of its container". It notes that "liquids and gases are *fluids*". Accordingly, the term *fluid dynamics* refers to the dynamic behavior of fluids, i.e. the behavior resulting from forces acting upon the fluids. For a fluid like water or air, typical origins for such forces are mechanical, for example: gravity, pressure gradients, or acceleration.

To describe the fluid's behavior, one assigns macroscopic quantities to the fluid, typically a density n, a velocity v, a pressure p, and a temperature T. A set of equations then describes how those quantities react to forces imposed on the fluid. These equations represent the conservation of mass, momentum, and energy. Additionally, an equation of state, for example the ideal gas equation is required.

For a plasma, the general idea stays the same. However, plasmas contain charged particles that react to mechanical as well as electromagnetic forces. Accordingly, additional force terms caused by electromagnetic fields need to be included, resulting in a special version of the Navier-Stokes equations. Also, since the particles create fields on their own, Maxwell's equations have to be included for a complete description. The combination is called *magnetohydrodynamics*. Usually the complete equations are simplified by further assumptions. See chapters 2 and 3 for more information about the equations used in this work.

The term *multi-fluid approach* refers to a model for a system with different kinds of particles. In particular for a plasma, there is at least one kind of ions, neutrals, and electrons, respectively. Each kind of particle (denoted by a subscript s) is modeled as an independent fluid with corresponding density n_s, velocity v_s, temperature T_s, and (partial) pressure p_s.

Fluid models are surprisingly successful for the simulation of plasmas, considering the number of simplifications and not always justified assumptions they make. Most likely, this can be explained by the inherent observance of the convervation laws, which are very strong principles in physics. However, results from fluid dynamic models are not always correct. Fluid dynamic models fail and thus are not suitable when certain assumptions inherent to fluid dynamics (e.g., thermodynamic equilibrium) do not hold anymore. Examples are especially common for plasmas near bounding walls, which are known for the occurrence of non-equilibrium phenomena. A model based on kinetic theory is necessary in that case. This kind of model will be described in section 1.3.

A good indicator as to use a fluid dynamic or a kinetic approach is the Knudsen number $Kn = \lambda/L$. It relates the mean free path of the particles in the fluid (i.e. the distance between two collisions) to the length of the observed system. For $Kn \ll 1$, a fluid model can be used, but once $Kn \gtrsim 0.1$ one should refer to kinetic theory. The following two examples can illustrate this: Assume a model for the complete discharge region of an HID lamp, so that $L \approx 1$ mm. The mean free path between heavy particles for typical discharge parameters is $\lambda \approx 10\,\mu$m. Accordingly, $Kn = \lambda/L \approx 10^{-2} \ll 1$ and a fluid dynamic model can be justified. When focusing on the near-cathode region only, $L \approx 10\,\mu$m, and $\lambda \approx 10\,\mu$m. Accordingly, $Kn \approx 1$ and fluid dynamic models should be expected to exhibit problems.

More information about fluid dynamic plasma modeling can be found, for example, in references [18]-[22].

1.3. Kinetic Modeling

The idea behind kinetic models is straightforward. The basic assumption is that one knows the state of every single particle at a given time, defined by its position and momentum. All forces from particle interactions can be calculated from this information. With additionally known external forces, the exact behavior of every particle and thus the system can be calculated from very basic equations. Unfortunately, this approach is doomed to fail, for the following reason: To remember the state of a single point in time requires the storage of approximately $36 \cdot 10^{23}$ values (position and momentum in three coordinates for about $6 \cdot 10^{23}$ particles). Assuming a typical 32 bit computing precision, the amount of data calculates to $1.34 \cdot 10^{16}$ GB. Storing those data on double-layer Blue-Ray discs (~ 50 GB capacity, 1.2 mm height) would result in a tower of $3.2 \cdot 10^8$ km height, approximately twice the distance between earth and sun. No present or foreseeable computer system is able to handle a problem this big.

Fortunately, such a high level of detail is not necessary. It suffices to consider the statistical behavior of the particles, so that the actual position and momentum of every particle can be represented by a $6 + 1$-dimensional probability distribution function $f_s(x, y, z, p_x, p_y, p_z, t)$ for each species s. Ludwig Boltzmann was the main contributor to this approach (cf. [23] and [24]) and the equation describing the behavior of the distribution function f_s was named after him.

The Boltzmann equation is – like the basic equations in the fluid dynamic model – a conservation equation. The actual conservation laws of mass, momentum, and energy can be calculated as weighted means of the Boltzmann equation, so-called moments. Analogously, macroscopic quantities like the density n_s or the mean velocity v_s can be calculated as moments of the distribution function f_s. The Boltzmann equation itself and its moments are further discussed in chapter 4.

A problem with the calculation of one moment is that it requires knowledge of the next higher moment. This results in a never-ending chain of equations. To be able to solve a model, one therefore has to artificially end that chain at some point. For example, the fluid dynamic equation system can be derived from kinetic theory by ending the chain after the second moment and replacing the third moment by an equation of state.

More information about kinetic plasma modeling can be found, for example, in references [18]-[22].

2. Investigation of a Common Fluid Model[1]

This chapter studies a multi-fluid model for the near-cathode region of a thermal plasma first presented by Benilov in 1995 [25]. The work investigates the model characteristics in dependence on α, which describes the ratio of the ionization length to the mean free path for ion-neutral collisions. Accordingly, $\alpha > 1$ implies that ion-neutral collisions (diffusion) determine the plasma behavior. Values of $\alpha < 1$ on the other hand imply that neutral-electron collisions (ionization) are dominant.

In [25], solutions could only be found for $\alpha \gg 1$. The model was modified in 1998 by Benilov and Naidis [26], where the authors assumed the edge of the Saha plasma to be reached due to full ionization, instead of an equilibrium between ionization and three body recombination. Solutions could be found for $\alpha \gg 1$ and $\alpha \ll 1$. For the case where $\alpha \approx \mathcal{O}(1)$, an approximation formula was presented. More recent publications treat fully ionized plasmas and neglect three body recombination [27, 28]. Despite these severe simplifications, they do not present a complete solution, either. This lack of a complete solution motivates a closer investigation, especially since comparisons of experimental data with the incomplete solution show reasonable results.

The purpose of this chapter is to determine whether the problems in finding a complete solution are due to the mathematical methods applied in [27, 28], or due to the design of the model. Some of the methods applied in the original paper, such as switching branches of square roots, are error-prone and require careful treatment. This chapter reinvestigates the fluid model by more straightforward methods, which are mathematically more robust.

To this aim, section 2.1 briefly introduces the physical model including the governing equations, as found in [28]. Subsequently, these equations are solved and the results and their interpretations are presented in sections 2.2 and 2.3.

---

[1]F H Scharf and R P Brinkmann, *J. Phys. D: Appl. Phys.* **39** (2006) 2738-2746

2.1. The Fluid Dynamic Model

2.1.1. Physical Structure

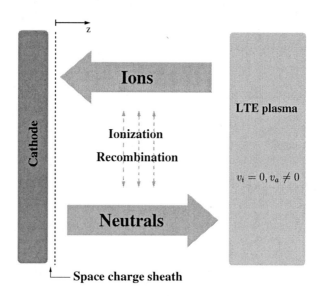

Figure 2.1.: Schematic structure of the near-cathode region. The z-axis is directed from left to right, with the beginning of the space charge sheath at $z = 0$. Recombination is neglected in the original fluid model.

The model proposed here assumes a stationary situation, so that variables like density and velocity do not change with time. The axial expansion of the modeled region (discussed in more detail further below) is very small compared with its radial expansion. This allows one to neglect radial variations and to employ a one-dimensional model along the z-axis.

This one-dimensional plasma region can be divided into several subregions with different characteristics, see for example [14], [29], and [30]. This work distinguishes three subregions: A space charge sheath, a presheath, and a fully ionized Saha plasma containing electrons and singly charged ions only (cf. figure 2.1). The Saha plasma is described by the Saha equation, which has been discussed in detail elsewhere, e.g., in chapter 2 of [31]. Electron and heavy particle temperatures (T_e and T_h) in the Saha plasma are assumed to be the same as in the presheath, but not necessarily equal. The space charge sheath is characterized by a depletion of electrons. The beginning of the space charge sheath is defined by the Bohm criterion, which defines the sheath edge as the point where the approximation of quasi-neutrality breaks down. It can be shown that the ion speed at this point is given by:

$$v_B = \sqrt{\frac{T_e}{m_i}} \tag{2.1}$$

More details about the Bohm criterion and sheath formation can be found in [32] and references therein, in particular [33].

The main focus of the model presented in this work is on the presheath, where ionization and acceleration occur. The space charge sheath and the Saha plasma represent boundary conditions to the presheath and thus to the model. The Saha boundary condition is usually not completely met for real discharges, the plasma continually develops into the plasma column instead (although on a larger scale than the one considered here, see below). However, choosing a Saha plasma as a boundary condition allows comparison with other, commonly used models, such as [28] and [14].

Along with these three regions, one finds three distinct length scales. (i) the Debye length λ_D, (ii) the mean free path for charge exchange collisions λ_{ia}, and (iii) the mean free path for ionization collisions λ_i. The smallest length is λ_D, measuring a few ten nanometers due to the very high electron densities in a typical discharge (around 10^{22} m^{-3}). λ_{ia} and λ_i typically measure a few ten micrometers and above. They are therefore much larger than λ_D, but no further assumption about the ratio between them is made. Accordingly, the lengths fulfill the relation

$$\lambda_i, \lambda_{ia} \gg \lambda_D. \tag{2.2}$$

This means that differences between the ion and electron density exist only in the space charge sheath, and quasi-neutrality ($n_i = n_e$) can be assumed for the presheath and the Saha plasma. Contrary to [14] and [30], the condition $\lambda_i \gg \lambda_{ia}$ is not assumed, and a further subdivision of the presheath into ionization layer and Knudsen layer is not made in this work. A rather detailed scale study including some numerical estimates can be found in [3].

With these assumptions, the multi-scale method allows one to neglect the expansion of the space charge sheath for the modeling of the presheath, so that the presheath begins at $z = 0$. Following also from the multi-scale method, the right boundary of the presheath is stretched to $z \to \infty$.

Between these two boundary conditions, one species of singly charged ions and the corresponding atoms engage in ionization and charge exchange collisions. Recombination is neglected at this point, which also leads to $v_a(z \to \infty) \neq 0$ and $n_{ap} = n_a(z \to \infty) = 0$. All heavy particles share a constant temperature T_h, and the electron temperature is constant at T_e. A variable temperature would improve the results, but at the price of increased complexity. Some recent fluid dynamic approaches make the extra effort to include a variable temperature (e.g., [34]), they may be able to allow a better estimate of how severe the simplification of a constant temperature is. The ratio $T_e/T_h = \beta$ is constant, and a value of $T_e/T_h = 1$ is used for most calculations in this work. This approximative value for β is supported by recent experimental results ([35] and [36]), where T_e/T_h varies between $1 \ldots 1.8$. The total pressure p is assumed to be constant and can thus be identified with the total pressure in the Saha plasma, where $n_{ap} = 0$:

$$
\begin{aligned}
p &= n_{ip}T_h + n_{ap}T_h + n_{ep}T_e \\
&= n_{ip}T_h + n_{ip}T_e \\
&= n_{ip}T_h(1 + \beta)
\end{aligned}
\tag{2.3}
$$

For this system one can now formulate one set of equations for each kind of heavy particle, i.e., one set for the neutrals and another set for the ions.

2.1.2. Equations

Neutral Atoms

The first equation is the continuity equation of the neutrals. It only includes the loss of neutrals due to ionization since recombination is neglected at this point. The edge of the Saha plasma is accordingly defined by full ionization, rather than an equilibrium between ionization and recombination. A constant ionization rate k_i is assumed:

$$
\frac{\mathrm{d}}{\mathrm{d}z}(n_a(z)v_a(z)) = -k_i n_a(z)n_e(z)
\tag{2.4}
$$

The second equation describes the momentum balance. Whenever a charge exchange collision occurs, the momentum of the neutral atom (ion) involved in the collision is transferred to the ion (atom) fluid. The effective momentum transfer thus depends on the relative velocities between the colliding particles. This is represented by the first term on the right. The second term on the right represents the momentum transfer due to ionization, which depends only on the speed of the neutrals:

$$\frac{d}{dz}(n_a(z)m_a v_a(z)^2 + T_h n_a(z)) = \frac{n_i(z)n_a(z)T_h^2}{pD_{ia}}(v_i(z) - v_a(z)) - k_i m_i n_i(z)n_a(z)v_a(z) \quad (2.5)$$

$D_{ia} = 2T_h/(m_i \nu_{ia})$ is the ion-neutral species diffusion coefficient with the corresponding collision frequency ν_{ia}. More details can be found in [28], for simplicity the charge exchange collisions will be written in terms of a collision rate $k_{cx} = T_h^2(m_i pD_{ia})^{-1}$ from here on.

Ions

A similar description yields the equations for the ions. The momentum transfer terms now appear with the opposite signs, because momentum lost by the neutral fluid is gained by the ion fluid, and vice versa. Since ions carry a charge, electromagnetic fields also need to be included. Due to the one-dimensional approach however, it suffices to include only the electric field:

$$\frac{d}{dz}(n_i(z)v_i(z)) = k_i n_a(z)n_e(z) \quad (2.6)$$

$$\frac{d}{dz}(m_i n_i(z)v_i(z)^2 + T_h n_i(z)) = en_i(z)E(z) - k_{cx}m_i n_i(z)n_a(z)(v_i(z) - v_a(z))$$
$$+ k_i m_i n_i(z)n_a(z)v_a(z) \quad (2.7)$$

Electrons

The electrons are assumed to be in Boltzmann equilibrium:

$$\frac{d}{dz}(n_e(z)T_e) = -en_e(z)E(z) \quad (2.8)$$

Elastic collisions between electrons and heavy particles are neglected. Given a constant electron temperature T_e, together with the condition of quasi-neutrality,

$$n_i(z) = n_e(z) \quad (2.9)$$

this yields a formula for the electric field:

11

$$E(z) = -\frac{T_e}{en_i(z)}\frac{\mathrm{d}n_i(z)}{\mathrm{d}z} \tag{2.10}$$

Since magnetic fields are not included in the model, equation 2.10 can be used instead of the complete system of Maxwell's equations.

2.2. Solution and Results

Altogether, the equations read:

$$\frac{\mathrm{d}}{\mathrm{d}z}(n_i v_i) = k_i n_i n_a, \tag{2.11}$$

$$\frac{\mathrm{d}}{\mathrm{d}z}(n_a v_a) = -k_i n_i n_a, \tag{2.12}$$

$$\frac{\mathrm{d}}{\mathrm{d}z}(n_i m_i v_i^2) = -(T_h + T_e)\frac{\mathrm{d}n_i}{\mathrm{d}z} - k_{cx}m_i n_i n_a(v_i - v_a) + k_i n_i n_a m_i v_a, \tag{2.13}$$

$$\frac{\mathrm{d}}{\mathrm{d}z}(n_a m_a v_a^2) = -T_h\frac{\mathrm{d}n_a}{\mathrm{d}z} + k_{cx}m_i n_i n_a(v_i - v_a) - k_i n_i n_a m_i v_a, \tag{2.14}$$

where the explicit dependence of the variables on z has been omitted for readability. The above is a set of four non-linear differential equations for four variables. Adding equations (2.11) and (2.12), as well as equations (2.13) and (2.14) yields – after integration – two algebraic expressions:

$$n_i v_i + n_a v_a = \text{const} \tag{2.15}$$

$$n_i m_i v_i^2 + n_a m_a v_a^2 = -(T_h + T_e)n_i - T_h n_a + \text{const} \tag{2.16}$$

Equation (2.15) represents the particle flux balance. Assuming that there is no net particle gain or loss within the ionization layer, the first constant is equal to zero. Equation (2.16) represents the pressure balance. By considering $z \to \infty$ in equation (2.16), which is equivalent to $n_a \to 0$ and $v_i \to 0$, the integration constant can be identified as the global pressure p.

Next, the equations are normalized. The base of normalization for z is the mean free path for ion-neutral collisions λ_{ia}, which are formally equivalent to charge exchange collisions. For the velocities and densities the bases are the neutral atom speed of sound (v_{as}) and the total heavy particle density in the Saha plasma ($n_{tp} = n_{ip} + n_{ap} = n_{ip}$), respectively. Since the Saha plasma is fully ionized, the total density is equal to the ion density:

$$z \to \lambda_{ia} z \qquad\qquad n \to n_{tp} n \qquad\qquad v \to v_{as} v$$
$$\lambda_{ia} = (n_{ip} k_i)^{-1} \sqrt{T_h/m_i} \quad n_{tp} = n_{ip} = p(T_h(1+\beta))^{-1} \quad v_{as} = \sqrt{T_h/m_i}$$
$$\alpha^2 = k_{cx}/k_i \qquad\qquad \beta = T_e/T_h$$

These definitions of α and β agree with the ones in [28]. All variables are now dimensionless, the Bohm speed is equal to $v_B = \sqrt{1 + T_e/T_h}$, and the neutral atom speed of sound is equal to $v_a = 1$. Also, from here the distinction between m_i and m_a will be dropped, since $m_i \approx m_a$ and $m_e \approx 0$.

Equations (2.15) and (2.16) are solved for n_i and n_a. The result can then be used to eliminate n_i and n_a in equations (2.13) and (2.14), yielding two algebraic equations for n_i and n_a as well as two differential equations for v_i and v_a:

$$n_i = \frac{(1+\beta)v_a}{v_a(1+\beta+v_i^2) - v_i - v_a^2 v_i} \tag{2.17}$$

$$n_a = -\frac{(1+\beta)v_i}{v_a(1+\beta+v_i^2) - v_i - v_a^2 v_i} \tag{2.18}$$

$$\frac{\mathrm{d}v_i}{\mathrm{d}z} = -\frac{v_i(1+\beta)(1+\beta - (1+\alpha^2)v_a v_i + (1+\alpha^2)v_i^2)}{(v_i^2 - 1 - \beta)(v_i + v_a^2 v_i - v_a(1+\beta+v_i^2))} \tag{2.19}$$

$$\frac{\mathrm{d}v_a}{\mathrm{d}z} = \frac{v_a(1+\beta)(\alpha^2 v_a(v_a - v_i) - 1)}{(v_a^2 - 1)(v_i + v_a^2 v_i - v_a(1+\beta+v_i^2))} \tag{2.20}$$

The parameter α is equivalent to the ratio of the mean free path for neutral-electron collisions (ionization) to the mean free path for ion-neutral collisions (diffusion). It can be understood as the inverse Knudsen number (see section 1.2) of the atoms or ions. Large values of α imply that collisions between ions and neutral atoms occur more frequently than collisions between neutrals and electrons, resulting in a diffusive behavior. Small values imply that ionization collisions are dominant. For $\alpha = 1$, the frequency of ionization collisions is equal to the frequency of charge exchange collisions. Typical values for α vary over a wide range, as demonstrated in [25]. α will be the varying parameter for the following examination of the system's behavior.

The parameter β is the ratio of the electron temperature to the heavy particle temperature. $\beta = 1$ is a common approximation (see [14] and [28]). As mentioned above, recent experiments ([35] and [36]) found β to be in the range between 1 and 2. Increasing β, however, was found to have little effect on the simulation results. Thus, the approximation $\beta = 1$ will be maintained.

The two differential equations (2.19) and (2.20) are non-linear and cannot be solved in a straightforward manner. However, since they contain only two variables, other methods can be applied to analyze the system analytically, e.g. plotting v_a over v_i. To simplify the calculations, a new parameter s is introduced, related to z by:

$$\frac{\mathrm{d}z}{\mathrm{d}s} = -\frac{(v_a^2 - 1)(v_i^2 - \beta - 1)(v_i + v_a^2 v_i - v_a(1 + v_i^2 + \beta))}{1 + \beta}$$

This parameter allows a mathematical description of the system without the problematic singularities of $\mathrm{d}v_a/\mathrm{d}v_i$ that stem from $\mathrm{d}v_i/\mathrm{d}z$ or $\mathrm{d}v_a/\mathrm{d}z$ becoming zero or infinite. The new equations for v_i and v_a read as follows:

$$\frac{\mathrm{d}v_i}{\mathrm{d}s} = -(v_a^2 - 1)v_i((v_a - v_i)v_i(1 + \alpha^2) - \beta - 1) \tag{2.21}$$

$$\frac{\mathrm{d}v_a}{\mathrm{d}s} = -v_a(v_a(v_a - v_i)\alpha^2 - 1)(v_i^2 - \beta - 1) \tag{2.22}$$

Integrating equations (2.21) and (2.22) with various initial values for v_i and v_a produces plots that help to determine solutions to the equations given above. One of these plots (for $\alpha = 2, \beta = 1$) is shown in figure 2.2. The plots show a clear symmetry, which together with the assertion that v_i and v_a must have a different sign (because ions and neutrals move in opposing directions), allows one to focus on either the upper left or the lower right quadrant. For the following investigations, the upper left quadrant will be used, as indicated by the white square in figure 2.2.

The behavior of the solutions depends on the value of α and can be divided into three cases, which can be distinguished as follows. Figure 2.2 contains the lines where $\mathrm{d}v_i/\mathrm{d}s$ or $\mathrm{d}v_a/\mathrm{d}s$ become zero.

$\mathrm{d}v_i/\mathrm{d}s = 0$:

$$v_i = 0 \tag{2.23}$$
$$v_a = \pm 1 \tag{2.24}$$
$$v_a = v_i + \frac{1}{v_i}\left(\frac{1 + \beta}{1 + \alpha^2}\right) \tag{2.25}$$

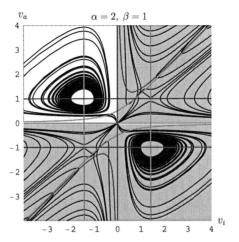

Figure 2.2.: Plot of exemplary solutions. The light gray (green and red for the color version) lines represent points where $\mathrm{d}v_a/\mathrm{d}v_i$ becomes zero or infinite. The white square defines the region of further investigation.

$\mathrm{d}v_a/\mathrm{d}s = 0$:

$$v_a = 0 \tag{2.26}$$

$$v_i = \pm\sqrt{1+\beta} \tag{2.27}$$

$$v_i = v_a - \frac{1}{\alpha^2 v_a} \tag{2.28}$$

Depending upon where these lines intersect, a different behavior is observed, which motivates the separation of the solution into different cases. Whereas equations (2.23), (2.24), (2.26), and (2.27) represent straight lines, equations (2.25) and (2.28) represent curves that are functions of v_i, v_a, and – most important – α. Their positions shift when α is varied.

The curve given by equation (2.25) lies completely in the two quadrants where v_i and v_a have the same sign. Only equation (2.28) yields a curve in the quadrant of focus. Investigating this curve and where it intersects with the straight lines (2.23) and (2.24) yields the cases sought. The curves (2.23) through (2.28) are therefore separatrices.

For the first case, the control parameter α is considered to be greater than unity. The curve then intersects with the line $v_i = 0$ below the line $v_a = 1$, defining the first case. When α decreases, the curve moves upwards. Below $\alpha = 1$, the intersection with $v_i = 0$ lies above $v_a = 1$, and there is a new intersection on the line $v_a = 1$, between $v_i = -\sqrt{1+\beta}$ and $v_i = 0$. This defines the second case. With α decreasing further, the intersection with the line $v_a = 1$ moves towards smaller v_i. The passing below the threshold of $v_i = -\sqrt{1+\beta}$ defines the third case; the value of α for this threshold is easily computed and will be denoted as α_{cr}:

$$\alpha_{\text{cr}} = \frac{1}{\sqrt{1 + \sqrt{1+\beta}}}$$

These three cases ($\alpha > 1$, $\alpha_{cr} < \alpha \le 1$, and $\alpha \le \alpha_{cr}$) cover all possible values of α. The following sections contain a detailed discussion of the existence of physical solutions in all three regions. A first requirement for a solution to be physical is that it starts with the Bohm criterion at the cathode ($v_i(z = 0) = -v_B = -\sqrt{1+\beta}$) and develops into a Saha plasma ($v_i(z \to \infty) = 0$). All other solutions will be disregarded immediately. The variables v_a, n_a, and n_i for the qualifying solutions will be calculated to further test these solutions for their physical validity. One important point in these tests will be the behavior of dz/ds. When dz/ds changes sign, z reverses its direction with respect to s. The solution at such a point turns and reverses its direction, too, comparable to the behavior of v_i at the Bohm point. In the quadrant of focus, there are two lines at which dz/ds changes sign: $v_i = -\sqrt{1+\beta}$ and $v_a = 1$. Both represent the Mach line for the corresponding species. Solutions traversing through these lines are generally non-physical. However, at points where the Mach lines intersect with other separatrices (e.g. intersection of (2.24) and (2.28)), a traversing, yet physical solution may be possible. This exception will be of special interest in section 2.2.2. Also, physical solutions may begin or end at the Mach lines.

2.2.1. Case I: $\alpha > 1$

All trajectories emerge from the singularity at $v_i = -\sqrt{1+\beta}, v_a = 1$. Each of them then spirals outwards in counter-clockwise direction. Only the tail of the spirals conforms to the boundary condition that v_i lie between $v_i = -\sqrt{1+\beta}$ and $v_i = 0$. All but one of the tails tend to the origin. Because crossing the origin implies a change of sign for the ion and atom velocity and thus represents a system with two cathodes and a plasma in between, such solutions are disregarded. This leaves only the solution given by the spiral's tail. Investigations show that this solution has all the properties sought: It covers the interval $v_i \in [-\sqrt{1+\beta}; 0]$, v_i increases with increasing z and the solution is defined on the interval $z \in [0; \infty[$. Since the solution does not traverse the lines $v_i = -\sqrt{1+\beta}$ or $v_a = 1$, it has a physical signification.

Some exemplary trajectories are shown in figure 2.3, with the solution highlighted by a bold line. The corresponding densities and velocities are shown in figures 2.4 and 2.5. There, the solid lines represent the density and velocity of the ions, respectively, whereas the dashed lines represent the neutral atom density and velocity, respectively. The neutral particle density exhibits a rapid decrease. Even though a true zero value is reached only for $z \rightarrow \infty$, the low neutral density raises concerns about the validity of the solution and the fluid description, which will be discussed in more detail later.

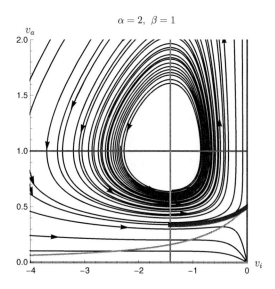

Figure 2.3.: Trajectories of v_i and v_a for $\alpha = 2$ including the lines where $\mathrm{d}v_a/\mathrm{d}v_i$ becomes zero or infinite. The unique solution is shown by a bold line.

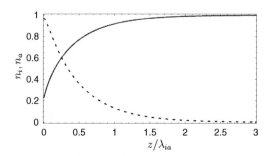

Figure 2.4.: Densities for the unique solution shown in figure 2.3. The solid line shows the ion density, the dashed line shows the neutral density ($\alpha = 2$).

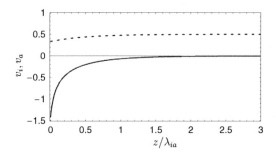

Figure 2.5.: Ion and neutral velocity for the unique solution. The solid line shows the ion velocity, the dashed line shows the neutral velocity ($\alpha = 2$).

2.2.2. Case II: $\alpha_{cr} < \alpha \leq 1$

Like in the case $\alpha > 1$, some of the trajectories again spiral outwards in counter-clockwise direction. The tail of exactly one spiral provides a solution that fulfills all of the necessary boundary conditions. This curve crosses the point $v_i = (\alpha^2 - 1)/\alpha^2$, $v_a = 1$. Closer investigations are necessary to determine whether this crossing yields a physical solution.

To this aim, v_i and v_a are linearized around the point $v_i^* = (\alpha^2 - 1)/\alpha^2$, $v_a^* = 1$:

$$
\begin{aligned}
v_i &= v_i^* + \delta v_i \\
v_a &= v_a^* + \delta v_a
\end{aligned}
$$

Since v_i^* and v_a^* are constants, the derivatives dv_i/ds and dv_a/ds are equivalent to $d\delta v_i/ds$ and $d\delta v_a/ds$. Written as a matrix $\underline{\underline{T}}$, the derivatives then read:

$$
\frac{d}{ds}\begin{pmatrix} v_i \\ v_a \end{pmatrix} = \frac{d}{ds}\begin{pmatrix} \delta v_i \\ \delta v_a \end{pmatrix} = \underline{\underline{T}}\begin{pmatrix} \delta v_i \\ \delta v_a \end{pmatrix}
\tag{2.29}
$$

The eigenvalues of $\underline{\underline{T}}$ determine the characteristics of the singularity. Linearizing dv_i/ds and dv_a/ds (equations (2.21) and (2.22)) yields:

$$
\underline{\underline{T}} = \begin{pmatrix} 0 & \dfrac{2(\alpha^2 - 1)(\alpha^4 + 1)}{\alpha^6} \\[2ex] -\dfrac{\alpha^4 + 2\alpha^2 - 1}{\alpha^2} & \dfrac{\alpha^6 + 3\alpha^4 + \alpha^2 - 1}{\alpha^4} \end{pmatrix}
\tag{2.30}
$$

The respective eigenvalues and eigenvectors are given by:

19

$$\lambda_1 = \frac{-1 + \alpha^2 + 3\alpha^4 + \alpha^6 + \sqrt{-7 + 22\alpha^2 - 21\alpha^4 + 20\alpha^6 + 3\alpha^8 - 2\alpha^{10} + \alpha^{12}}}{2\alpha^4} \quad (2.31)$$

$$\boldsymbol{v}_{\lambda 1} = \left(\frac{\alpha^2 \lambda_1}{\alpha^4 + 2\alpha^2 - 1}, 1 \right)^T \quad (2.32)$$

$$\lambda_2 = \frac{-1 + \alpha^2 + 3\alpha^4 + \alpha^6 - \sqrt{-7 + 22\alpha^2 - 21\alpha^4 + 20\alpha^6 + 3\alpha^8 - 2\alpha^{10} + \alpha^{12}}}{2\alpha^4} \quad (2.33)$$

$$\boldsymbol{v}_{\lambda 2} = \left(\frac{\alpha^2 \lambda_2}{\alpha^4 + 2\alpha^2 - 1}, 1 \right)^T \quad (2.34)$$

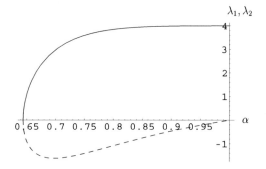

Figure 2.6.: Eigenvalues λ_1 (solid line) and λ_2 (dashed line) for α between α_{cr} and zero.

For $\alpha_{cr} < \alpha < 1$, λ_1 is always positive and λ_2 is always negative (cf. figure 2.6). The fixed point at $\{v_i^* = (\alpha^2 - 1)/\alpha^2, v_a^* = 1\}$ is therefore a saddle point. The solutions of the differential equation tend towards the fixed point in the case of the negative eigenvalue, and away from the fixed point in the case of the positive eigenvalue. These solutions are defined for the mathematical parameter s, the physical parameter is z. As shown before, z changes direction at $v_a = 1$, leading to two valid solutions that pass through the singularity in z-space (see figure 2.7).

A detailed investigation of the singularity leads to the following: In the limit $\alpha \to 1$, λ_2 becomes zero and the singularity becomes a terrace point, which represents a smooth transition to the behavior shown for $\alpha < 1$. The singularity degenerates for $\alpha \to \alpha_{cr}$, where both eigenvalues become zero, before it changes into a stable spiral point for $\alpha < \alpha_{cr}$ (see next section).

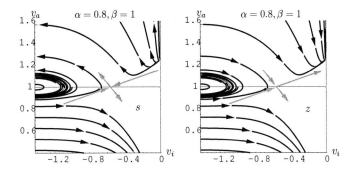

Figure 2.7.: Behavior of the solutions in vicinity of the singularity in s- and z-parametrization (left and right, respectively). Black lines represent the trajectories of v_i and v_a, with s increasing in direction of the arrows. Light gray (turquoise in the color version) arrows represent the eigenvectors corresponding to the calculated eigenvalues λ_1 and λ_2.

The eigenvectors corresponding to the calculated eigenvalues define two straight lines that cross the singularity; they are tangents to the solutions through the singularity and are therefore included in figures 2.7 and 2.8. For the region $\alpha_{cr} < \alpha \leq 1$, the slope of the first line is always positive and the slope of the second line is always negative. The solution that belongs to the tangent with negative slope can be disregarded, because it eventually crosses the origin; however, the first tangent belongs to a valid solution. In addition to this solution, there are other trajectories above the neutral particle Mach line $v_a = 1$, that fulfill the boundary conditions for the ions (Bohm criterion on one end, zero velocity on the other). However, for these solutions, the neutrals leave the space charge sheath already at supersonic speeds. Additionally, assuming that the solution should connect smoothly to the solutions found for $\alpha < 1$, the solution yielded by the spiral should be favored. It is therefore highlighted in figure 2.8.

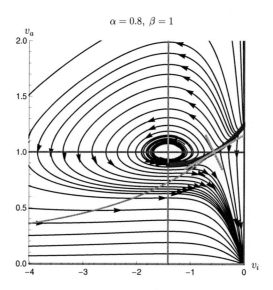

Figure 2.8.: Trajectories of v_i and v_a for $\alpha = 0.8$ including the lines where $\mathrm{d}v_a/\mathrm{d}v_i$ becomes zero or infinite. The favored solution is shown by a bold line; the tangents of the solution crossing the singularity are also included.

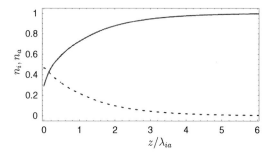

Figure 2.9.: Densities for the favored solution shown in plot 2.8. The solid line shows the ion density, the dashed line shows the neutral density ($\alpha = 0.8$).

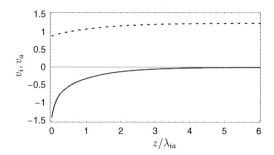

Figure 2.10.: Ion and neutral velocity for the favored solution. The solid line shows the ion velocity, the dashed line shows the neutral velocity ($\alpha = 0.8$).

2.2.3. Case III: $\alpha < \alpha_{cr}$

In the third case, all trajectories emerge from the point $v_i = 0, v_a = 1/\alpha$ and either circle into a new singularity to the left of $v_i = -\sqrt{1 + \beta}$ or cross the origin (see figure 2.11). All trajectories fulfill the boundary conditions and therefore represent a formal solution. All solutions show supersonic neutral velocities on the whole interval $z \in [0, \infty[$, thus their physical significance is highly questionable. No unique solution can be found. For demonstration purposes, one solution was picked, the corresponding densities and velocities are shown in figures 2.12 and 2.13.

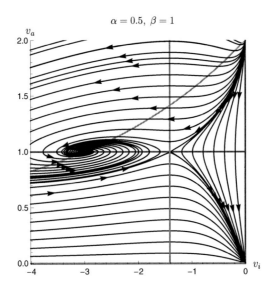

$\alpha = 0.5, \ \beta = 1$

Figure 2.11.: Trajectories of v_i and v_a for $\alpha = 0.5$ including the lines where $\mathrm{d}v_a/\mathrm{d}v_i$ becomes zero or infinite. No unique solution can be found.

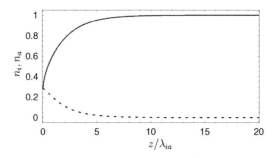

Figure 2.12.: Densities for an exemplary solution. The solid line shows the ion density, the dashed line shows the neutral density ($\alpha = 0.5$).

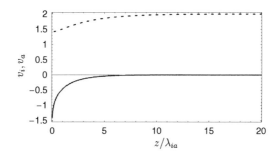

Figure 2.13.: Ion and neutral velocity for an exemplary solution. The solid line shows the ion velocity, the dashed line shows the neutral velocity ($\alpha = 0.5$).

2.3. Summary and Interpretation of the Results

A detailed analysis yielded three cases for which the solutions of the differential equations (2.19) and (2.20) show a specific characteristic behavior. The regions are defined by the intersections of separatrices and can be identified in dependence on α.

A unique solution can only be found for $\alpha \geq 1$. For $\alpha_{cr} < \alpha < 1$, there are several possible solutions. However, one solution shows a prominent behavior that suggests to choose it over the other solutions. All solutions however include supersonic neutrals. For $\alpha \leq \alpha_{cr}$, no unique solution can be found and for all solutions the neutrals are supersonic everywhere in the ionization layer.

Altogether, this reproduces the results reported in [28], especially for the case $\alpha > 1$. For $\alpha < 1$, the results differ slightly, mainly due to a different physical interpretation of the mathematical solutions. Since straightforward methods were used to achieve the results described here, the method of solving the equations or numerical errors are highly unlikely to cause the lack of a complete solution in [28]. Instead, the equations themselves must bear the problems. In particular the neglect of recombination and the fluid description itself are likely to cause the problems. To investigate this further, the next chapter will expand the fluid model to allow for recombination.

3. Expansion of the Fluid Model by Three Body Recombination[2]

The previous chapter has shown that the fluid model in its current form lacks physical validity and applicability. The neglect of three body recombination was identified as an important factor. In this section, the model will be expanded to include three body recombination. In particular, this means that the Saha plasma at $z \to \infty$ is not fully ionized anymore, and $n_{ap} \neq 0$. Since the particle flux balance (2.15) still needs to be fulfilled, this means that $v_a(z \to \infty) = 0$. In addition to the expansion by three body recombination, an improved method of implementing the boundary condition given by the Saha plasma at infinity will be presented.

3.1. The Expanded Model

3.1.1. Equations

The starting point for the formulation are the equations (2.4) - (2.7), which represent the equations of continuity and motion for ions and neutral atoms. They are now extended by adding a recombination term, $k_r n_i^3$, with k_r representing the three body recombination rate:

$$\frac{\mathrm{d}}{\mathrm{d}z}(n_i v_i) = k_i n_i n_a - k_r n_i^3 \tag{3.1}$$

$$\frac{\mathrm{d}}{\mathrm{d}z}(n_a v_a) = -k_i n_i n_a + k_r n_i^3 \tag{3.2}$$

$$\frac{\mathrm{d}}{\mathrm{d}z}(n_i m_i v_i^2) = -(T_h + T_e)\frac{\mathrm{d}n_i}{\mathrm{d}z} - n_i n_a k_{cx} m_i (v_i - v_a) + k_i n_i n_a m_i v_a - k_r n_i^3 m_i v_i \tag{3.3}$$

$$\frac{\mathrm{d}}{\mathrm{d}z}(n_a m_i v_a^2) = -T_h \frac{\mathrm{d}n_a}{\mathrm{d}z} + n_i n_a k_{cx} m_i (v_i - v_a) - k_i n_i n_a m_i v_a + k_r n_i^3 m_i v_i \tag{3.4}$$

---

[2]F H Scharf and R P Brinkmann, *J. Phys. D: Appl. Phys.* **41** (2008) 185206

3. Expansion of the Fluid Model by Three Body Recombination

It is assumed, in these equations, that the electron mass is negligible ($m_e \ll m_i$) and that the mass of the ions and neutrals are equal ($m_a \approx m_i$). k_i and k_{cx} are the rates of ionization and charge exchange collisions, respectively. The equations are completed by the conservation laws for mass and momentum. These conservation forms can be integrated to yield the flux and pressure balances. With no net particle gain or loss and a constant pressure p, they read:

$$n_i v_i + n_a v_a = \text{const.} = 0 \tag{3.5}$$

$$\begin{aligned} m_i n_i v_i^2 + m_i n_a v_a^2 &= -(T_h + T_e)n_i - T_h n_a + \text{const.} \\ &= -(T_h + T_e)n_i - T_h n_a + p \end{aligned} \tag{3.6}$$

As in the previous chapter, p can be identified with the total pressure in the Saha plasma:

$$\begin{aligned} p &= n_{ip}T_h + n_{ap}T_h + n_{ep}T_e \\ &= n_{ip}T_h + \gamma n_{ip}T_h + n_{ip}T_e \\ &= n_{ip}T_h(1 + \gamma + \beta) \end{aligned} \tag{3.7}$$

For further calculations, all equations are normalized by use of the following rules and normalization constants:

$$z \to \lambda_{ia} z \qquad\qquad n \to n_{tp} n \qquad\qquad v \to v_{as} v$$

$$\lambda_{ia} = (n_{ip}k_i)^{-1}\sqrt{T_h/m_i} \qquad n_{tp} = p(1+\gamma)(T_h(1+\beta+\gamma))^{-1} \qquad v_{as} = \sqrt{T_h/m_i}$$

$$\alpha^2 = k_{cx}/k_i \qquad\qquad \beta = T_e/T_h \qquad\qquad \gamma = n_{ap}/n_{ip} = n_{ip}k_r/k_i$$

In these normalization rules, v_{as} is the speed of sound for the neutral atoms. Accordingly, $v_a = 1$ represents the Mach line for the neutrals. The basis for the density normalization is the total heavy particle density in the plasma, $n_{tp} = n_{ip} + n_{ap} = n_{ip}(1 + \gamma)$. The basis for the length normalization is the mean free path for ion-atom collisions, λ_{ia}. The normalization constants represent the ratio between charge exchange and ionization collisions (α^2) between electron and heavy particle temperature (β) and between recombination and ionization (γ). The normalized equations of continuity and motion, plus the flux and pressure balance equations then read:

$$\frac{d}{dz}(n_i v_i) = (n_i n_a - \gamma(1+\gamma)n_i^3)(1+\gamma) \tag{3.8}$$

$$\frac{d}{dz}(n_a v_a) = (-n_i n_a + \gamma(1+\gamma)n_i^3)(1+\gamma) \tag{3.9}$$

$$\frac{d}{dz}(n_i v_i^2) = -(1+\beta)\frac{dn_i}{dz} - (\alpha^2 n_i n_a(v_i - v_a) - n_i n_a v_a + \gamma(1+\gamma)n_i^3 v_i)(1+\gamma) \tag{3.10}$$

$$\frac{d}{dz}(n_a v_a^2) = -\frac{dn_a}{dz} + (\alpha^2 n_i n_a(v_i - v_a) - n_i n_a v_a + \gamma(1+\gamma)n_i^3 v_i)(1+\gamma) \tag{3.11}$$

$$n_i v_i + n_a v_a = 0 \tag{3.12}$$

$$n_i v_i^2 + n_a v_a^2 = -(1+\beta)n_i - n_a + \frac{1+\beta+\gamma}{1+\gamma} \tag{3.13}$$

Proceeding with the methods presented in the previous chapter, two algebraic equations for the densities n_i and n_a as well as two differential equations for the velocities v_i and v_a can be formulated from the equation system above:

$$n_i = -\frac{v_a(\beta + \gamma + 1)}{(v_i v_a^2 - (v_i^2 + \beta + 1)\,v_a + v_i)\,(\gamma + 1)} \tag{3.14}$$

$$n_a = \frac{v_i(\beta + \gamma + 1)}{(v_i v_a^2 - (v_i^2 + \beta + 1)\,v_a + v_i)\,(\gamma + 1)} \tag{3.15}$$

Substituting these two equations for the densities in the ion equations of continuity (3.1) and motion (3.3), one can formulate a set of two non-linear differential equations for v_i and v_a. With the abbreviation $\beta_1 = \beta + 1$, they read:

$$\begin{aligned}
\frac{dv_i}{dz} = {}&-\frac{(\beta_1 + \gamma)}{(\beta_1 - v_i^2)(v_i + v_a^2 v_i - v_a(\beta_1 + v_i^2))^2} \\
&\times \Bigg[(1+\alpha^2)v_a^3 v_i^3 - v_i^2(\beta_1 + (1+\alpha^2)v_i^2) \\
&\quad + v_a^2(\beta_1 \gamma(\beta_1 + \gamma) - (2+\alpha^2)\beta_1 v_i^2 - 2(1+\alpha^2)v_i^4) \\
&\quad + v_a v_i(\beta_1^2 + (1 + 2\beta_1 + \alpha^2(1+\beta_1))v_i^2 + (1+\alpha^2)v_i^4) \Bigg]
\end{aligned} \tag{3.16}$$

$$\frac{dv_a}{dz} = \frac{(\beta_1 + \gamma)v_a}{(v_a^2 - 1)v_i(v_i + v_a^2 v_i - v_a(\beta_1 + v_i^2))^2}$$

$$\times \left[v_a^4(\gamma(\beta_1 + \gamma) + \alpha^2 v_i^2) - v_i^2 \right.$$

$$- v_a^3 v_i(\alpha^2 \beta_1 + \gamma(\beta_1 + \gamma) - 2\alpha^2 v_i^2)$$

$$+ v_a v_i(\beta_1 - (-1 + \alpha^2)v_i^2)$$

$$\left. + v_a^2(\gamma(\beta_1 + \gamma) + (-1 + \alpha^2(1 + \beta_1))v_i^2 + \alpha^2 v_i^4 \right] \qquad (3.17)$$

For the case when there is no recombination ($\gamma = 0$), these equations become equal to equations (2.19) and (2.20). They are too complex to solve analytically, but they can be solved numerically for $n_i, n_a, v_i,$ and v_a. The required boundary conditions are discussed next.

3.1.2. Boundary Conditions

To solve the differential equations (3.16) and (3.17) numerically, starting values for v_i and v_a are necessary. Such starting values need to be derived from physical boundary conditions. The model used here offers two possible conditions:

- Ions have (negative) Bohm velocity $v_i = -v_B = -\sqrt{1 + \beta}$ at $z = 0$

- Saha plasma at $z \to \infty$

The first condition can be implemented easily, but it yields only one boundary condition (i.e., for the ions) where two (for ions and neutrals) are needed. In [37], this was overcome by using a shooting method to find a boundary condition for the neutrals, so that the solutions tend towards a Saha plasma. Such methods are very time consuming and in this case cannot hit the real solution, but can only approach it. A different way to find appropriate starting values is desirable.

Such a way can be deducted from the second condition. It yields two equations, but cannot be applied directly to numerical calculations because the point at which the condition is given is infinity. Numerical calculations never reach infinity and require boundary conditions at finite distances. However, assuming a Saha plasma at any finite location only yields the trivial solution, which is a Saha plasma everywhere. It is therefore necessary to find boundary conditions that are defined at a finite location, but ensure transition into a Saha plasma as z grows to infinity. This can be done by linearizing equations (3.8) - (3.11) around the Saha plasma, defined by $\{n_i = n_{ip}, n_a = n_{ap}, v_i = 0, v_a = 0\}$, or normalized $\{n_i = 1/(1+\gamma), n_a = \gamma/(1+\gamma), v_i = 0, v_a = 0\}$. By adding a small perturbation to the equilibrium conditions, the behavior in the vicinity of the Saha plasma can be investigated:

$$n_i(z) = 1/(1+\gamma) + \delta n_i(z) \tag{3.18}$$
$$n_a(z) = \gamma/(1+\gamma) + \delta n_a(z) \tag{3.19}$$
$$v_i(z) = 0 + \delta v_i(z) \tag{3.20}$$
$$v_a(z) = 0 + \delta v_a(z) \tag{3.21}$$

Substituting definitions (3.18) - (3.21) into equations (3.8) - (3.11) and linearizing the result yields a simplified equation system for $\delta n_i(z), \delta n_a(z), \delta v_i(z)$, and $\delta v_a(z)$:

$$(\delta n_a(z) - 2\gamma\delta n_i(z))(1+\gamma) - \delta v_i'(z) = 0 \tag{3.22}$$
$$(2\gamma\delta n_i(z) - \delta n_a(z))(1+\gamma) - \gamma\delta v_a'(z) = 0 \tag{3.23}$$
$$\left(\alpha^2 + 1\right)\gamma(\delta v_a(z) - \delta v_i(z)) - (\beta+1)(\gamma+1)\delta n_i'(z) = 0 \tag{3.24}$$
$$\delta n_a'(z)(\gamma+1) + \left(\alpha^2 + 1\right)\gamma(\delta v_a(z) - \delta v_i(z)) = 0 \tag{3.25}$$

This equation system is simple enough to find an analytic solution containing four integration constants. However, three of those four constants can be calculated from the necessary condition that all perturbations need to vanish for $z \to \infty$. The remaining solutions contain only one constant:

$$\delta n_i(z) = -C_1 e^{-\lambda z} \tag{3.26}$$
$$\delta n_a(z) = C_1(1+\beta)e^{-\lambda z} \tag{3.27}$$
$$\delta v_i(z) = -C_1\frac{(1+\beta)\lambda}{1+\alpha^2}e^{-\lambda z} \tag{3.28}$$
$$\delta v_a(z) = C_1\frac{(1+\beta)\lambda}{\gamma(1+\alpha^2)}e^{-\lambda z} \tag{3.29}$$
$$\lambda = \sqrt{\frac{(1+\alpha^2)(1+\gamma)(1+\beta+2\gamma)}{1+\beta}} > 0 \tag{3.30}$$

The solutions exhibit an exponentially decreasing behavior which agrees with the idea of a distorted equilibrium. The integration constant C_1 is arbitrary, as long as it is larger than zero. It can be moved into the exponent and then represents a shift of the z-coordinate. By setting $C_1 = 1$, the shift vanishes and equations (3.26) - (3.29) provide a simple means to define starting values for n_i, n_a, v_i, and v_a. However, care has to be taken when γ becomes small, as both $|\delta v_i(z)|$ and $|\delta v_a(z)|$ then grow rapidly and the approximation of a small distortion does not hold anymore (cf. figure 3.1). This can be mended by moving to larger values of z or by using, e.g., $C_1 = \gamma^2$. The latter definition of C_1 puts a limit on the growth of $|\delta v_i(z)|$ and $|\delta v_a(z)|$, see figure 3.2.

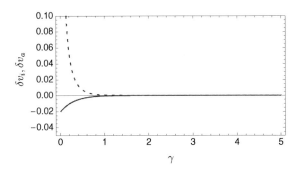

Figure 3.1.: Velocity perturbations δv_i (solid) and δv_a (dashed) from equations (3.28) and (3.29) for $\alpha = 1, \beta = 1, C_1 = 1, z = 3$, in dependence on the normalized recombination rate γ. For very small values of γ, δv_i and δv_a grow apart.

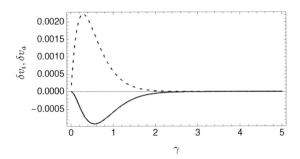

Figure 3.2.: Velocity perturbations δv_i (solid) and δv_a (dashed) from equations (3.28) and (3.29) for $\alpha = 1, \beta = 1, C_1 = 1, z = 3$, in dependence on the normalized recombination rate γ. $|\delta v_i - \delta v_a|$ now has a maximum at $\gamma \approx 1/2$.

3.2. Solutions

The solutions can be divided into two groups of different behavior. Before investigating this further, however, one example will be given to show that the new model can also reproduce results calculated with the original model.

3.2.1. Example I: $\{\alpha = 2, \beta = 1, \gamma = 10^{-9}\}$

This case is chosen to verify that the extended model can reproduce results produced with the original model. Since $\gamma = 0$ would reintroduce numerical issues and singularities, a very small recombination rate $\gamma = 10^{-9}$ is chosen instead. The results look identical to the ones obtained previously (compare figures 3.3 - 3.5 with figures 2.3 - 2.11), until rather large values of z are reached. The new velocities then approach zero due to the fact that γ is very small, but not exactly zero. This occurs outside of the z-range plotted in figure 3.5, but can be seen in figure 3.3.

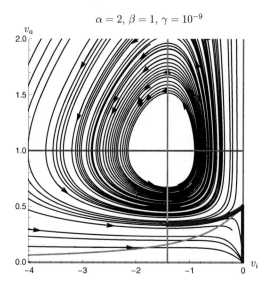

$$\alpha = 2, \ \beta = 1, \ \gamma = 10^{-9}$$

Figure 3.3.: Trajectories of v_i and v_a for $\{\alpha = 2, \beta = 1, \gamma = 10^{-9}\}$. The unique solution is shown by a bold line. The steep drop near $v_i = 0$ is caused by γ being very small but not exactly zero. The light gray (or red and green in the color version) lines represent points where $\mathrm{d}v_a/\mathrm{d}v_i$ becomes infinite or zero.

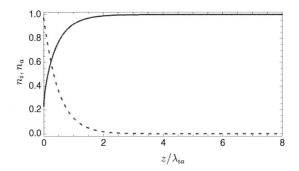

Figure 3.4.: Densities n_i (solid) and n_a (dashed) for the unique solution in figure 3.3. The sudden drop observed there for the velocities is out of the plot range ($\{\alpha = 2, \beta = 1, \gamma = 10^{-9}\}$).

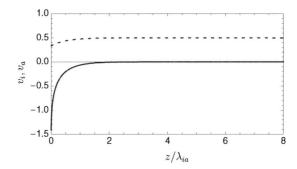

Figure 3.5.: Velocities v_i (solid) and v_a (dashed) for the unique solution in figure 3.3. The sudden drop observed there for the velocities is out of the plot range ($\{\alpha = 2, \beta = 1, \gamma = 10^{-9}\}$).

3. Expansion of the Fluid Model by Three Body Recombination

Next, the general behavior of the solutions in dependence on α, β, and γ is investigated. As mentioned above, the behavior can be divided into two groups, mostly defined by α and γ. β slightly influences the quantitative results, but has no influence on the qualitative results.

The first group consists of solutions which do not yield any supersonic atom velocities. The exact relationship defining the borderline to the second group is very complex and an analytic expression could not be found. Figure 3.6 shows the borderline in the α-γ-space for $\beta = 1, 2$, and 3. Whenever the parameter combination $\{\alpha, \gamma\}$ for a given β is is on the left side of the curve in figure 3.6, the neutral atom velocity will rise above the speed of sound before the ions have reached Bohm velocity. (This will be described below.) On the other hand, no supersonic velocities occur when the combination of $\{\alpha, \gamma\}$ is on the right side.

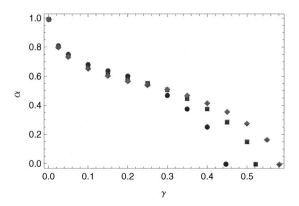

Figure 3.6.: Borderline for occurrence of supersonic neutrals in the α-γ-space, for $\beta = 1$ (circles), $\beta = 2$ (squares), and $\beta = 3$ (diamonds). Combinations of $\{\alpha, \gamma\}$ left from the borderline lead to supersonic neutrals, combinations to the right do not.

The top left and bottom right points of the borderline yield special values for α and γ, from here on referred to as α_{min} and γ_{min}. If α and/or γ are larger than their respective special value, the velocities are always sub-sonic, even if the other parameter becomes zero. For the top left point ($\gamma \to 0$), one finds $\alpha_{min} = 1$, which agrees with the findings chapter 2, where $\gamma = 0$. For the bottom right point ($\alpha \to 0$), the corresponding γ_{min} was calculated by numerical experiments. The results can be interpolated extremely well by $\gamma_{min}(\beta) = \beta^{1/4}/\sqrt{5}$, as shown in figure 3.7.

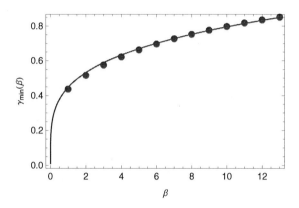

Figure 3.7.: Minimum values γ_{min} to ensure that no supersonic neutrals occur when $\alpha \to 0$, plotted in dependence on β at $\beta = 1 \ldots 13$ (circles) and the interpolation function $\gamma_{min}(\beta) = \beta^{1/4}/\sqrt{5}$ (continuous line).

Examples II and III further illustrate the behavior of solutions within the first group.

3.2.2. Example II: $\{\alpha = 2, \beta = 1, \gamma = 1\}$

This example assumes that ion and neutral density in the Saha plasma are equal, which can be observed clearly in figure 3.9. As expected, the recombination rate leads to a slow-down of the neutral atoms; their speed tends toward zero in the Saha plasma. Since $\gamma = 1 > \gamma_{\min}$, no supersonic velocities occur (cf. figure 3.10). Shape and approximate values of all curves agree very well with similar calculations in [37].

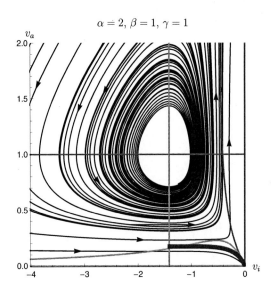

Figure 3.8.: Trajectories of v_i and v_a. The unique solution is shown by a bold line ($\{\alpha = 2, \beta = 1, \gamma = 1\}$). The light gray (or red and green in the color version) lines represent points where dv_a/dv_i becomes infinite or zero.

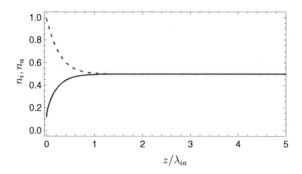

Figure 3.9.: Densities n_i (solid) and n_a (dashed) for the unique solution in figure 3.8 ($\{\alpha = 2, \beta = 1, \gamma = 1\}$).

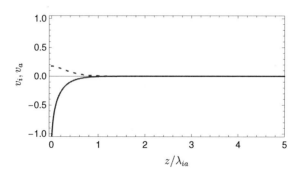

Figure 3.10.: Velocities v_i (solid) and v_a (dashed) for the unique solution in figure 3.8 ($\{\alpha = 2, \beta = 1, \gamma = 1\}$).

3.2.3. Example III: $\{\alpha = 0.4, \beta = 1, \gamma = 0.4\}$

γ is chosen below the minimal value of $\gamma_{\mathrm{min}}(\beta = 1) \approx 1/\sqrt{5} \approx 0.447$, and α is chosen above the minimal value $\alpha_{\mathrm{min}}(1, 0.4) \approx 0.25$, but below the critical $\alpha_{\mathrm{cr}}(\beta) = 1/\sqrt{1 + \sqrt{1 + \beta}}$ found in chapter 2, which computes to approximately 0.64 for $\beta = 1$. As expected, the expanded model yields a good result without supersonic neutrals, whereas the original model without three body recombination could not yield a result at all for $\alpha < \alpha_{\mathrm{cr}}$ (figures 3.11 - 3.13 vs. figures 2.11 - 2.13).

When both α and γ are below their respective minimal values, the neutral velocity increases to the speed of sound before the ions have reached the Bohm speed and no physical solution exists. This can be best explained by example IV.

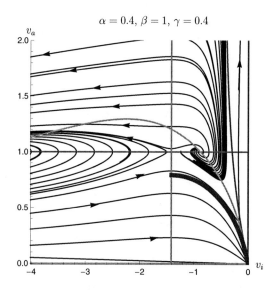

Figure 3.11.: Trajectories of v_i and v_a. The unique solution is shown by a bold line ($\{\alpha = 0.4, \beta = 1, \gamma = 0.4\}$). The light gray (or red and green in the color version) lines represent points where $\mathrm{d}v_a/\mathrm{d}v_i$ becomes infinite or zero.

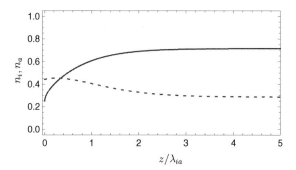

Figure 3.12.: Densities n_i (solid) and n_a (dashed) for the unique solution in figure 3.11 ($\{\alpha = 0.4, \beta = 1, \gamma = 0.4\}$).

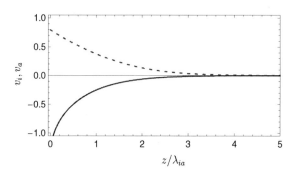

Figure 3.13.: Velocities v_i (solid) and v_a (dashed) for the unique solution in figure 3.11 ($\{\alpha = 0.4, \beta = 1, \gamma = 0.4\}$).

3.2.4. Example IV: $\{\alpha = 0.1, \beta = 1, \gamma = 0.4\}$

Figures 3.14 - 3.16 show the behavior of all solutions in the v_i-v_a-space, as well as the densities and velocities for the one solution emerging from a Saha plasma. This specific solution ends when the neutrals reach the speed of sound. The line that seems to continue the solution in figure 3.14 cannot be connected to the solution because the direction of z is inverted at $v_a = 1$. Following that line above $v_a = 1$ would therefore increase the (spatial) distance from the sheath edge again. Thus, there exists no solution that fulfills the Bohm criterion on one end and the Saha equation on the other end.

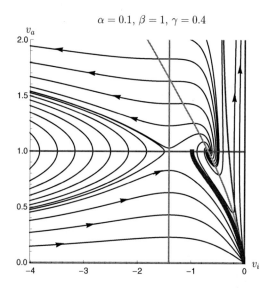

Figure 3.14.: Trajectories of v_i and v_a. The only possible, but unphysical solution is shown by a bold line ($\{\alpha = 0.1, \beta = 1, \gamma = 0.4\}$). The light gray (or red and green in the color version) lines represent points where dv_a/dv_i becomes infinite or zero.

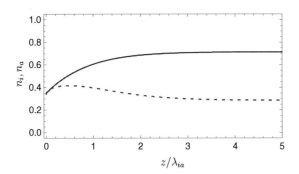

Figure 3.15.: Densities n_i (solid) and n_a (dashed) for the only possible, but unphysical solution in figure 3.14 ($\{\alpha = 0.1, \beta = 1, \gamma = 0.4\}$).

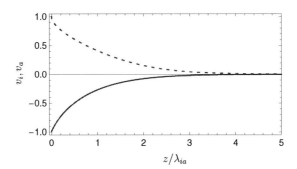

Figure 3.16.: Velocities v_i (solid) and v_a (dashed) for the only possible, but unphysical solution in figure 3.14 ($\{\alpha = 0.1, \beta = 1, \gamma = 0.4\}$).

3.3. Summary and Interpretation of the Results

The original multi-fluid model of the near-cathode region for one species of atom and one species of singly charged ion with a constant temperature was expanded to allow for three body recombination. The expanded model features ionization, recombination, charge exchange collisions, and forces from a self-consistently calculated electric field. In order to solve the model, a method to numerically implement the physical boundary conditions imposed by the Saha plasma at infinity was presented.

It was shown that the solutions to the model can be divided into two groups of distinct behaviors in dependence on the two parameters α (representing charge exchange collisions) and γ (representing three body recombination). The general relation between α and γ, defining the borderline between the two groups, could not be formulated analytically, but numerically calculated plots for varying β were shown (figure 3.6). Additionally, minimum values for α (or γ) that ensure subsonic velocities for $\gamma \to 0$ (or $\alpha \to 0$) were given.

In the first group, the solutions generally agree with solutions from chapter 2 and other publications, specifically [28] and [37]. The first group, however, also contains solutions for parameter values for which no solutions were possible in those publications, specifically $\alpha = 0.4$, as in example III. All solutions of the first group show a transition from a Saha plasma to a Bohm sheath edge and no supersonic atoms.

In the second group, all solutions emerging from a Saha plasma feature atom speeds that reach supersonic values before the ions have reached the Bohm speed. The solutions also end at that point due to the structure of the model equations; they cannot be continued beyond the neutral Mach line. Accordingly, there exist no solutions connecting a Saha plasma to a Bohm sheath edge for parameter combinations of α and γ within the second group.

For the case where α and γ are exactly on the borderline between the two groups, solutions that cross the Mach line are also possible. However, this case is unstable and has no physical relevance, since the slightest perturbation of α or γ eliminates that possibility.

In summary, the multi-fluid model presented in this chapter extends the abilities of established multi-fluid models. Like those models however, the new model is still not able to yield solutions for all physical parameters. To further improve the quality of results and access the now still inaccessible parameter regions, it is necessary to switch to kinetic methods, which will be looked at in the next chapter.

4. Construction of a Kinetic Model

The previous two chapters have shown that the capabilities of fluid dynamics to describe the near-cathode region in thermal plasmas are limited. The limitations manifested themselves in the appearance of singularities and unphysical solutions, effectively limiting the range of allowed values for the available model parameters. The following two chapters present a kinetic model and its solution to lift the limitations of fluid dynamics. In a way, the presented model is similar to earlier works by Schmitz and Riemann ([29] and [30]). Schmitz and Riemann present a kinetic model for the near-cathode region in thermal plasmas which also employs the multi-scale method and boundary conditions similar to the ones used here. However, the authors assume that the neutrals are Maxwell-distributed with a constant temperature T_h and that the ion dynamics are dominated by symmetric charge exchange collisions with constant frequency ν_{cx}. The model proposed here assumes a more general form of the neutral distribution function. Additionally, while taking charge exchange collisions into account, it describes the interaction between ions and neutrals by a more generic Boltzmann collision term.

4.1. The Kinetic Model

The starting point for the kinetic model is the Boltzmann equation, which describes the interaction between the particles:

$$\frac{\partial f(t, \boldsymbol{r}, \boldsymbol{v})}{\partial t} + \boldsymbol{v} \cdot \frac{\partial f(t, \boldsymbol{r}, \boldsymbol{v})}{\partial \boldsymbol{r}} + \dot{\boldsymbol{v}} \cdot \frac{\partial f(t, \boldsymbol{r}, \boldsymbol{v})}{\partial \boldsymbol{v}} = < f(t, \boldsymbol{r}, \boldsymbol{v}) >_c \qquad (4.1)$$

This equation and the corresponding collision term $< f(t, \boldsymbol{r}, \boldsymbol{v}) >_c$ need to be discussed separately for each kind of particle. The kinetic model discussed here includes three kinds of particles: One species of singly charged ions (subscript 'i', mass m_i), one species of neutral atoms (subscript 'a', $m_a \approx m_i$), and electrons (subscript 'e', $m_e \approx 0$). Accordingly, the model employs three distribution functions. The heavy particle distribution functions $f_a(t, \boldsymbol{r}, \boldsymbol{v})$ and $f_i(t, \boldsymbol{r}, \boldsymbol{v})$ are of special interest, since their moments represent physical quantities like the heavy particle density, flux, energy, or kinetic temperature (which is equivalent to the measurable temperature in thermodynamic equilibrium):

$$n_{i/a}(t, \boldsymbol{r}) = \int f_{i/a}(t, \boldsymbol{r}, \boldsymbol{v}) \mathrm{d}^3 v$$

$$\boldsymbol{\Gamma}_{i/a}(t, \boldsymbol{r}) = n_{i/a}(t, \boldsymbol{r})\boldsymbol{u}_{i/a}(t, \boldsymbol{r}) = \int \boldsymbol{v} f_{i/a}(t, \boldsymbol{r}, \boldsymbol{v}) \mathrm{d}^3 v$$

$$e_{i/a}(t, \boldsymbol{r}) = \int \frac{m_i \boldsymbol{v}^2}{2} f_{i/a}(t, \boldsymbol{r}, \boldsymbol{v}) \mathrm{d}^3 v$$

$$T_{i/a}(t, \boldsymbol{r}) = \int m_i (\boldsymbol{v} - \boldsymbol{u}_{i/a})^2 f_{i/a}(t, \boldsymbol{r}, \boldsymbol{v}) \mathrm{d}^3 v$$

Knowledge of these macroscopic quantities is necessary to calculate the transfer functions q_p and j.

The Boltzmann equation is very complex, especially for non-trivial right-hand sides. To be able to solve it, it is usually necessary to introduce a number of simplifications and make assumptions on the structure of the distribution functions. In this work, the heavy particles are assumed to be in thermal equilibrium in the Saha plasma (at $z \to \infty$), where they share a common temperature T_h. Accordingly, their distribution function at $z \to \infty$ can be assumed to be Maxwellian. With increasing distance from the Saha plasma (i.e., towards smaller values of z), the particles build up a drift velocity that increases to the order of the thermal velocity. This leads to a shift and a distortion of the distribution function in phase space. The distribution functions for the heavy particles can therefore be expected to be disturbed Maxwellian distributions. Although generally no assumption is made about how strong the disturbances are, the convergence of the model can be expected to decrease with increasing size of the disturbances.

In contrast to the heavy particles, the electrons are usually so far from thermal equilibrium, that no suitable assumption can be made about the electron distribution function. Instead, similar to the fluid model, one assumes that the electrons are in Boltzmann equilibrium:

$$n_e \sim e^{\frac{e\Phi}{T_e}} \tag{4.2}$$

It states that the electron density is close to the density of a Maxwell distribution with temperature T_e. Since the assumption is not made about the distribution function itself but the density (the first moment of the distribution function) instead, it is a weaker assumption then, for example, the assumption of thermodynamic equilibrium. By use of $\boldsymbol{E} = -\nabla \Phi$, equation (4.2) can be used to relate the electron density to the electric field, as shown in equation (2.8). Further, by introducing rate coefficients for collisions involving electrons, one can determine the heavy particle distribution functions without having to solve the Boltzmann equations for the electrons explicitly.

An additional simplification is the assumption of a stationary situation. All solutions are time-independent and all time derivatives $\partial/\partial t$ are equal to zero. The definition of the distribution functions can be simplified accordingly: $f(t, r, v) = f(r, v)$.

The same model structure that was assumed for the fluid models (cf. figure 2.1 in section 2.1.1) is assumed for the kinetic model, which means that the multi-scale approach holds. Furthermore, the one-dimensional approximation suffices for the spatial coordinates. In velocity space, radial components cannot be completely neglected because that would lead to faulty collision terms. However, cylindric symmetry can be assumed. The distribution function can then be written as $f(r, v) = f(z, v) = f(z, v_\parallel, v_\perp^2)$. Although this is already a shorter definition than the original $f(t, r, v)$, including the arguments of f every time makes the formulas verbose and difficult to read. To avoid this, arguments will only be listed when they are non-standard (i.e. different from z, v) or when a special emphasis is desired. The following forms can therefore be used interchangeably: $f = f(z, v) = f(v_\parallel, v_\perp^2) = f(z, v_\parallel, v_\perp^2) = f(v)$. Finally, effects of gravity are neglected for all particles.

4.2. Equations

4.2.1. Kinetic Equation for the Ions

With time-derivatives neglected, equation (4.1) reads:

$$v \cdot \frac{\partial f_i}{\partial r} + \dot{v} \cdot \frac{\partial f_i}{\partial v} = < f_i >_c \qquad (4.3)$$

$\dot{v}$ is the ion acceleration and can be written as Eq_i/m_i, with ion mass m_i and ion charge $q_i = Ze$. In this work, only singly charged ions are considered, so that $Z = 1$.

A more detailed look at the collision term $< f_i >_c$ is now necessary. The following collision terms are taken into account:

I Ionization

II Recombination

III Elastic collisions among heavy particles

IV Heat sink representing radial thermal diffusion

Ionization refers to inelastic collisions between neutrals and electrons. No further assumption about the electron distribution is necessary at this point besides that one can define an electron density $n_e(z)$. With this density, ionization can be described using a (constant) ionization rate coefficent k_i, multiplied by the neutral distribution function. Ionization adds particles to the ions, the corresponding term therefore carries a positive sign:

$$< f_i >_{cI} = +k_i n_e(z) f_a(z, \boldsymbol{v}) \qquad (4.4)$$

Recombination is the counterpart to ionization. The most important kind of recombination, for the examined regime, is three body recombination involving two electrons and one ion; it can be described using a recombination rate coefficient k_r and electron density $n_e(z)$ (squared because two electrons are needed):

$$< f_i >_{cII} = -k_r n_e(z)^2 f_i(z, \boldsymbol{v}) \qquad (4.5)$$

Additionally, elastic collisions between ions and neutrals are taken into account. Charge exchange collisions are included in the elastic collisions since the result of both collisions cannot be distinguished. Elastic collisions are very important to realize an equilibrium between the particles. They can be described by a Boltzmann collision term, which describes the collision of two particles (indices 1 and 2). Each particle is described by its distribution function, $f_1(t, \boldsymbol{r_1}, \boldsymbol{v})$ and $f_2(t, \boldsymbol{r}, \tilde{\boldsymbol{v}})$, respectively. The dependence on t and $\boldsymbol{r}$ can be neglected at this point, thus the functions become $f_1(\boldsymbol{v})$ and $f_2(\tilde{\boldsymbol{v}})$. When the particles collide, their velocities change from $(\boldsymbol{v}, \tilde{\boldsymbol{v}})$ to $(\boldsymbol{v}', \tilde{\boldsymbol{v}}')$. The probability for a collision between the two particles is closely related to the collision cross section (e.g., [19]) and generally depends on the velocity of the particles. For simplicity, a constant cross section W_{12} will be assumed in this work. The situation is symmetric, meaning that the probability for the inverse collision $(\boldsymbol{v}', \tilde{\boldsymbol{v}}') \to (\boldsymbol{v}, \tilde{\boldsymbol{v}})$ is the same: $W_{12} = W_{21} = W$. The effect of a single collision on the distribution function of particle 1 then consists of a component that adds to the distribution function and the inverse component, which is subtracted from the distribution function:

$$\mathrm{d} < f_1(\boldsymbol{v}) >_{el} = W \left(f_1(\boldsymbol{v}') f_2(\tilde{\boldsymbol{v}}') - f_1(\boldsymbol{v}) f_2(\tilde{\boldsymbol{v}}) \right) \qquad (4.6)$$

To calculate the complete collision term, one has to take into account collisions at all possible speeds. This is simply done by integrating (4.6):

$$< f_1(\boldsymbol{v}) >_{el} = < f_1 >_{el} = \int_{\tilde{\boldsymbol{v}}} \int_{\tilde{\boldsymbol{v}}'} \int_{\boldsymbol{v}'} W \left(f_1(\boldsymbol{v}') f_2(\tilde{\boldsymbol{v}}') - f_1(\boldsymbol{v}) f_2(\tilde{\boldsymbol{v}}) \right) \mathrm{d}\boldsymbol{v}' \mathrm{d}\tilde{\boldsymbol{v}}' \mathrm{d}\tilde{\boldsymbol{v}} \qquad (4.7)$$

It is important to note that only certain combinations of $v, \tilde{v}, v'$, and $\tilde{v}'$ are valid, namely those that satisfy the conservation laws of mass and momentum. An elegant way to mathematically implement this restriction is shown below, following the approach presented in [38]. Before, however, it is helpful to rephrase (4.7) in center-of-mass coordinates

Rephrasing into Center-of-Mass Coordinates

Any given set of particles has a center of mass. The movement of the center can be described as a weighted average of all particles. The movement of any single particle can then be described in relation to the movement of the center and every other particle. For the case of only two particles, the respective formulas become rather simple.

$$c = \frac{m_1 v + m_2 \tilde{v}}{m_1 + m_2} \tag{4.8}$$

$$c' = \frac{m_1 v' + m_2 \tilde{v}'}{m_1 + m_2} \tag{4.9}$$

$$g = v - \tilde{v} \tag{4.10}$$

$$g' = v' - \tilde{v}' \tag{4.11}$$

g and g' are the difference velocities between particles 1 and 2. With these definitions, $v, \tilde{v}, v'$, and $\tilde{v}'$ can be rewritten as follows:

$$v = c + \frac{m_2}{m_1 + m_2} g \tag{4.12}$$

$$\tilde{v} = c - \frac{m_1}{m_1 + m_2} g \tag{4.13}$$

$$v' = c' + \frac{m_2}{m_1 + m_2} g' \tag{4.14}$$

$$\tilde{v}' = c' - \frac{m_1}{m_1 + m_2} g' \tag{4.15}$$

This can be substituted in equation (4.7). The new variables for the integrals are g, c', and g'. It can be shown that the Jacobian for this variable transformation is unity and the new integral reads

4. Construction of a Kinetic Model

$$
\begin{aligned}
< f_1 >_c = \int_g \int_{c'} \int_{g'} W \Bigg(& f_1(c' + \frac{m_2}{m_1 + m_2}g')f_2(c' - \frac{m_1}{m_1 + m_2}g') \\
& - f_1(c + \frac{m_2}{m_1 + m_2}g)f_2(c - \frac{m_1}{m_1 + m_2}g) \Bigg) \, \mathrm{d}g'\mathrm{d}c'\mathrm{d}g.
\end{aligned}
\tag{4.16}
$$

In this notation, the conservation laws of momentum and energy are easily implemented, as shown in the next section.

Conservation of Momentum and Energy

The conservation of momentum in Cartesian coordinates reads

$$
m_1 v + m_2 \tilde{v} = m_1 v' + m_2 \tilde{v}'. \tag{4.17}
$$

Dividing both sides by $(m_1 + m_2)$ gives the corresponding equation in center-of-mass coordinates:

$$
\frac{m_1 v + m_2 \tilde{v}}{m_1 + m_2} = \frac{m_1 v' + m_2 \tilde{v}'}{m_1 + m_2} \tag{4.18}
$$
$$
\Rightarrow c = c'
$$

The movement of the center is not affected by collisions between the two particles. This can be proven for any number of particles, but for the purpose of this work two particles are sufficient.

The conservation of energy reads

$$
\frac{m_1 v^2}{2} + \frac{m_2 \tilde{v}^2}{2} = \frac{m_1 v'^2}{2} + \frac{m_2 \tilde{v}'^2}{2}. \tag{4.19}
$$

Substituting equations (4.12) - (4.15) in equation (4.19) and reordering yields a very simple relation to describe the energy conservation law:

$$
\frac{g^2}{2} = \frac{g'^2}{2} \tag{4.20}
$$

Due to their simple form, both laws can be mathematically implemented as delta distributions, thus ensuring the physical validity of the collision term:

$$
< f_1 >_c = \int_g \int_{c'} \int_{g'} W \delta \left(\frac{g^2}{2} - \frac{g'^2}{2} \right) \delta^{(3)}(c - c')
$$
$$
\times \left(f_1(c' + \frac{m_2}{m_1 + m_2} g') f_2(c' - \frac{m_1}{m_1 + m_2} g') \right.
$$
$$
\left. - f_1(c + \frac{m_2}{m_1 + m_2} g) f_2(c - \frac{m_1}{m_1 + m_2} g) \right) dg' dc' dg.
$$

The integration over c' can now be carried out directly and applying the recursion formula $\delta(x^2 - a^2) = (2|a|)^{-1}[\delta(x - a) + \delta(x + a)]$ simplifies the term even further:

$$
< f_1 >_c = \int_g \int_{g'} \frac{W}{g} \left(\delta(g - g') + \delta(g + g') \right)
$$
$$
\times \left(f_1(c + \frac{m_2}{m_1 + m_2} g') f_2(c - \frac{m_1}{m_1 + m_2} g') \right. \tag{4.21}
$$
$$
\left. - f_1(c + \frac{m_2}{m_1 + m_2} g) f_2(c - \frac{m_1}{m_1 + m_2} g) \right) dg' dg
$$

In spherical coordinates, $dg = g^2 \sin(\theta) dg d\theta d\varphi$ and $dg' = g'^2 \sin(\theta') dg' d\theta' d\varphi'$. The corresponding limits for the integration are:

- $[0, \infty)$ for g and g'

- $[0, \pi)$ for θ and θ'

- $[0, 2\pi)$ for φ and φ'

Before integrating over g, it is necessary to substitute

$$
c = v - \frac{m_2}{m_1 + m_2} g = v - \frac{m_2}{m_1 + m_2} g e
$$

into equation (4.21), where e represents a unit vector that points into the direction of g. Accordingly, $g' = g' e'$. Also, it is convenient to introduce

$$\mu_{1/2} = \frac{m_{1/2}}{m_1 + m_2}.$$

Then, the integral over g can be carried out directly. With $d\Omega = \sin(\theta)d\theta d\varphi$ and $d\Omega' = \sin(\theta')d\theta' d\varphi'$, the resulting integral reads:

$$< f_1 >_c = \int_{\Omega'} \int_{\Omega} \int_{g'} W g'^3 \left(f_1(\boldsymbol{v} - \mu_2 g'\boldsymbol{e} + \mu_2 g'\boldsymbol{e}') f_2(\boldsymbol{v} - \mu_2 g'\boldsymbol{e} - \mu_1 g'\boldsymbol{e}') \right.$$
$$\left. - f_1(\boldsymbol{v}) f_2(\boldsymbol{v} - (\mu_2 + \mu_1)g'\boldsymbol{e}) \right) dg' d\Omega d\Omega' \tag{4.22}$$

This result is used to formulate one collision term for ion-ion collisions (probability W_{ii}) and one for ion-atom collisions (probability W_{ia}). In both cases, $\mu_1 = \mu_2 = 1/2$, since $m_a \approx m_i$. The complete collision term for elastic collisions is obtained by summation:

$$< f_i >_{cIII} = \int_{\Omega'} \int_{\Omega} \int_{g'} \left[W_{ii} g'^3 \left(f_i \left(\boldsymbol{v} - \frac{g'\boldsymbol{e}}{2} + \frac{g'\boldsymbol{e}'}{2} \right) f_i \left(\boldsymbol{v} - \frac{g'\boldsymbol{e}}{2} - \frac{g'\boldsymbol{e}'}{2} \right) \right. \right.$$
$$\left. - f_i(\boldsymbol{v}) f_i(\boldsymbol{v} - g'\boldsymbol{e}') \right)$$
$$+ W_{ia} g'^3 \left(f_i \left(\boldsymbol{v} - \frac{g'\boldsymbol{e}}{2} + \frac{g'\boldsymbol{e}'}{2} \right) f_a \left(\boldsymbol{v} - \frac{g'\boldsymbol{e}}{2} - \frac{g'\boldsymbol{e}'}{2} \right) \right.$$
$$\left. \left. - f_i(\boldsymbol{v}) f_a(\boldsymbol{v} - g'\boldsymbol{e}') \right) \right] dg' d\Omega d\Omega' \tag{4.23}$$

One other collision term is included, which accounts for the energy that normally diffuses radially. Since $\partial/\partial R$ is set to zero (one-dimensional approach), this radial diffusion is neglected in this model. This leads to a steady non-physical energy flux towards the cathode. To counterbalance that, a kind of thermal sink is introduced artificially for both ions and neutrals. It consists of collisions with pool of very light ($\ll m_i$) virtual particles, represented by a Maxwellian distribution with temperature T_h. Mathematically, the corresponding collision term can be deduced by a Taylor expansion of the Boltzmann collision term:

$$< f_i >_{cIV} = \nu_1 \left(3f_i + \boldsymbol{v} \cdot \nabla_v f_i + \frac{T_h}{m_i} \nabla_v^2 f_i \right) \tag{4.24}$$

With definitions (4.4)-(4.24) and the corresponding simplifications, equation (4.3) can be reduced to the following form:

$$v_\parallel \frac{\partial f_i}{\partial z} + \frac{E_z q_i}{m_i} \frac{\partial f_i}{\partial v_\parallel} = <f_i>_c = <f_i>_{cI} + <f_i>_{cII} + <f_i>_{cIII} + <f_i>_{cIV} \quad (4.25)$$

If the electric field $\boldsymbol{E}$ is to be determined self-consistently, the equation becomes more complicated. By assumption of quasi-neutrality ($n_i \approx n_e$) within the sheath, the electric field can be written as follows:

$$\boldsymbol{E} = -\nabla\Phi = -\frac{T_e}{e n_e(\boldsymbol{r})}\nabla n_e(\boldsymbol{r}) = -\frac{T_e}{e n_i(\boldsymbol{r})}\nabla n_i(\boldsymbol{r}) \quad (4.26)$$

Combining all collision terms and also assuming quasi-neutrality $n_e \approx n_i$ for the collision terms leads to

$$
\begin{aligned}
&v_\parallel \frac{\partial f_i}{\partial z} - \frac{T_e}{m_i n_i}\frac{\mathrm{d}n_i}{\mathrm{d}z}\frac{\partial f_i}{\partial v_\parallel} \\
&= k_i n_i f_a - k_r n_i^2 f_i + \nu_1\left(3f_i + \boldsymbol{v}\cdot\nabla_{\boldsymbol{v}} f_i + \frac{T_h}{m_i}\nabla_{\boldsymbol{v}}^2 f_i\right) \\
&\quad + \int_{\Omega'}\int_\Omega\int_{g'} g'^3\left[W_{ii}\left(f_i\left(\boldsymbol{v} - \frac{g'\boldsymbol{e}}{2} + \frac{g'\boldsymbol{e}'}{2}\right) f_i\left(\boldsymbol{v} - \frac{g'\boldsymbol{e}}{2} - \frac{g'\boldsymbol{e}'}{2}\right) - f_i(\boldsymbol{v})f_i(\boldsymbol{v} - g'\boldsymbol{e}')\right)\right. \\
&\qquad \left. + W_{ia}\left(f_i\left(\boldsymbol{v} - \frac{g'\boldsymbol{e}}{2} + \frac{g'\boldsymbol{e}'}{2}\right) f_a\left(\boldsymbol{v} - \frac{g'\boldsymbol{e}}{2} - \frac{g'\boldsymbol{e}'}{2}\right) - f_i(\boldsymbol{v})f_a(\boldsymbol{v} - g'\boldsymbol{e}')\right)\right] \\
&\quad \times \mathrm{d}g'\mathrm{d}\Omega\mathrm{d}\Omega'
\end{aligned}
$$

$$(4.27)$$

4.2.2. Kinetic Equation for the Neutral Atoms

The neutral atoms can be described similarly to the ions. Since they do not react directly to the electric field, the equivalent to equation (4.3) simply reads

$$\boldsymbol{v}\cdot\frac{\partial f_a}{\partial \boldsymbol{r}} = <f_a>_c . \quad (4.28)$$

The corresponding collision terms have the same structure as for the ions, but carry the opposite sign for the inelastic collisions:

$$< f_a >_{cI} = -k_i n_e(z) f_a(z, \boldsymbol{v}) \tag{4.29}$$

$$< f_a >_{cII} = +k_r n_e(z)^2 f_i(z, \boldsymbol{v}) \tag{4.30}$$

$$
\begin{aligned}
< f_a >_{cIII} = \int_{\Omega'} \int_{\Omega} \int_{g'} & \Bigg[W_{aa} g'^3 \left(f_a \left(\boldsymbol{v} - \frac{g' \boldsymbol{e}}{2} + \frac{g' \boldsymbol{e}'}{2} \right) f_a \left(\boldsymbol{v} - \frac{g' \boldsymbol{e}}{2} - \frac{g' \boldsymbol{e}'}{2} \right) \right. \\
& \left. - f_a(\boldsymbol{v}) f_a(\boldsymbol{v} - g' \boldsymbol{e}') \right) \\
& + W_{ai} g'^3 \left(f_a \left(\boldsymbol{v} - \frac{g' \boldsymbol{e}}{2} + \frac{g' \boldsymbol{e}'}{2} \right) f_i \left(\boldsymbol{v} - \frac{g' \boldsymbol{e}}{2} - \frac{g' \boldsymbol{e}'}{2} \right) \right. \\
& \left. - f_a(\boldsymbol{v}) f_i(\boldsymbol{v} - g' \boldsymbol{e}') \right) \Bigg] \mathrm{d}g' \mathrm{d}\Omega \mathrm{d}\Omega' \tag{4.31}
\end{aligned}
$$

$$< f_a >_{cIV} = \nu_1 \left(3 f_a + \boldsymbol{v} \cdot \nabla_{\boldsymbol{v}} f_a + \frac{T_h}{m_i} \nabla_{\boldsymbol{v}}^2 f_a \right) \tag{4.32}$$

This leads to the following form for the neutral particle Boltzmann equation, again assuming quasi-neutrality:

$$
\begin{aligned}
v_{\parallel} \frac{\partial f_a}{\partial z} & \\
= -k_i n_i f_a & + k_r n_i^2 f_i + \nu_1 \left(3 f_a + \boldsymbol{v} \cdot \nabla_{\boldsymbol{v}} f_a + \frac{T_h}{m_i} \nabla_{\boldsymbol{v}}^2 f_a \right) \\
+ \int_{\Omega'} \int_{\Omega} \int_{g'} g'^3 & \Bigg[W_{aa} \left(f_a \left(\boldsymbol{v} - \frac{g' \boldsymbol{e}}{2} + \frac{g' \boldsymbol{e}'}{2} \right) f_a \left(\boldsymbol{v} - \frac{g' \boldsymbol{e}}{2} - \frac{g' \boldsymbol{e}'}{2} \right) - f_a(\boldsymbol{v}) f_a(\boldsymbol{v} - g' \boldsymbol{e}') \right) \\
& + W_{ai} \left(f_a \left(\boldsymbol{v} - \frac{g' \boldsymbol{e}}{2} + \frac{g' \boldsymbol{e}'}{2} \right) f_i \left(\boldsymbol{v} - \frac{g' \boldsymbol{e}}{2} - \frac{g' \boldsymbol{e}'}{2} \right) - f_a(\boldsymbol{v}) f_i(\boldsymbol{v} - g' \boldsymbol{e}') \right) \Bigg] \\
& \times \mathrm{d}g' \mathrm{d}\Omega \mathrm{d}\Omega'
\end{aligned}
$$

$$\tag{4.33}$$

4.3. Boundary Conditions and Conservation Laws

The final solutions for f_i and f_a have to fulfill a set of boundary conditions. The following sections will identify these conditions and investigate how they can be fulfilled.

4.3.1. Saha Plasma

At the far end of the sheath ($z \to \infty$), there is a Saha plasma. Accordingly, the distribution functions have to assume a Maxwellian form for $z \to \infty$:

$$f_{i/aM}(v^2) = f_{i/a}(z, v)|_{z \to \infty} = n_{ip/ap} \left(\frac{m_i}{2\pi T_h} \right)^{3/2} \exp \left(-\frac{m_i v^2}{2T_h} \right) \qquad (4.34)$$

As a result, the left side of the kinetic equations (4.3) and (4.28) is zero and the collision term should also be zero. It can easily be proven that this holds for the terms (4.4) - (4.24).

To this aim, the ionization and the recombination part of the collision terms are considered first. Inserting equation (4.34) for the distribution function gives the necessary relation between ionization and recombination frequency, so that both effects counterbalance each other in the Saha plasma:

$$< f_i >_{cI} |_{z \to \infty} + < f_i >_{cII} |_{z \to \infty} = 0$$
$$\Rightarrow \quad k_i n_i(z) f_a(z, v)|_{z \to \infty} - k_r n_i(z)^2 f_i(z, v)|_{z \to \infty} = 0$$
$$\Rightarrow \quad k_i n_{ip} n_{ap} \left(\frac{m_i}{2\pi T_h} \right)^{\frac{3}{2}} \exp \left(-\frac{m_i v^2}{2T_h} \right) = k_r n_{ip}^2 n_{ip} \left(\frac{m_i}{2\pi T_h} \right)^{\frac{3}{2}} \exp \left(-\frac{m_i v^2}{2T_h} \right)$$
$$\Rightarrow \quad k_i n_{ip} n_{ap} = k_r n_{ip}^3$$
$$\Rightarrow \quad \frac{n_{ap}}{n_{ip}^2} = \frac{k_r}{k_i}$$

Second, the effect of the elastic collisions needs to be investigated, which takes a little more effort than the other terms. In the following equations, f_1 and f_2 are the distribution functions of two particles where '1' and '2' represent any arbitrary combination of ions and atoms. $W_{12} = W_{21}$ is the mutual cross section.

$$< f >_{cIII} |_{z \to \infty} = \int_{\Omega'} \int_{\Omega} \int_{g'} W_{12} g'^3 \left(f_1 \left(\boldsymbol{v} - \frac{g'e}{2} + \frac{g'e'}{2} \right) f_2 \left(\boldsymbol{v} - \frac{g'e}{2} - \frac{g'e'}{2} \right) \right.$$

$$\left. - f_1(\boldsymbol{v}) f_2(\boldsymbol{v} - g'e') \right) |_{z \to \infty} \mathrm{d}g' \mathrm{d}\Omega \mathrm{d}\Omega'$$

$$= \int_{\Omega'} \int_{\Omega} \int_{g'} W_{12} g'^3 n_{1p} n_{2p} \left(\frac{m_i}{2\pi T_h} \right)^3$$

$$\times \left[\exp \left(-\frac{m_i}{2\pi T_h} \left(\left(\boldsymbol{v} - \frac{g'e}{2} + \frac{g'e'}{2} \right)^2 + \left(\boldsymbol{v} - \frac{g'e}{2} - \frac{g'e'}{2} \right)^2 \right) \right) \right.$$

$$\left. - \exp \left(-\frac{m_i}{2\pi T_h} \left(\boldsymbol{v}^2 + (\boldsymbol{v} - g'e')^2 \right) \right) \right] \mathrm{d}g' \mathrm{d}\Omega \mathrm{d}\Omega'$$

$$= \int_{\Omega'} \int_{\Omega} \int_{g'} W_{12} g'^3 n_{1p} n_{2p} \left(\frac{m_i}{2\pi T_h} \right)^3$$

$$\times \left[\exp \left(-\frac{m_i}{2\pi T_h} \left(2\boldsymbol{v}^2 - 2g'\boldsymbol{v} \cdot \boldsymbol{e} + g'^2 \right) \right) \right.$$

$$\left. - \exp \left(-\frac{m_i}{2\pi T_h} \left(2\boldsymbol{v}^2 - 2g'\boldsymbol{v} \cdot \boldsymbol{e} + g'^2 \right) \right) \right] \mathrm{d}g' \mathrm{d}\Omega \mathrm{d}\Omega'$$

$$= 0$$

The term for elastic collisions between any arbitrary combination of ions and atoms vanishes, which means that the sum of those terms also vanishes. Finally, the fourth collision term has to be evaluated at $z \to \infty$. This can be done straightforwardly, as shown below for the ion term. The calculation for the corresponding neutral term can be done analogously.

$$< f_i >_{cIV} |_{z \to \infty} = \nu_1 \left(3 f_i(z, \boldsymbol{v})|_{z \to \infty} + \boldsymbol{v} \cdot \nabla_{\boldsymbol{v}} f_i(z, \boldsymbol{v})|_{z \to \infty} + \frac{T_h}{m_i} \nabla_{\boldsymbol{v}}^2 f_i(z, \boldsymbol{v})|_{z \to \infty} \right)$$

$$= \nu_1 \left(3 f_{iM}(v^2) + v \frac{\mathrm{d} f_{iM}(v^2)}{\mathrm{d}v} + \frac{T_h}{m_i} \frac{1}{v^2} \frac{\mathrm{d}}{\mathrm{d}v} \left(v^2 \frac{\mathrm{d} f_{iM}(v^2)}{\mathrm{d}v} \right) \right)$$

$$= \nu_1 \left(3 f_{iM}(v^2) - \frac{m_i}{T_h} v^2 f_{iM}(v^2) + \frac{T_h}{m_i} \frac{1}{v^2} \frac{\mathrm{d}}{\mathrm{d}v} \left(-v^2 \frac{m_i}{T_h} v f_{iM}(v^2) \right) \right)$$

$$= \nu_1 \left(3 f_{iM}(v^2) - \frac{m_i}{T_h} v^2 f_{iM}(v^2) + \frac{1}{v^2} \left(\frac{m_i}{T_h} v^4 f_{iM}(v^2) - 3v^2 f_{iM}(v^2) \right) \right)$$

$$= 0$$

This concludes the proof that the model satisfies the Saha plasma boundary condition as long as $k_i n_{ap} = k_r n_{ip}^2$. It should be noted that this condition relates n_{ip} and n_{ap} to each other, but it does not require a specific value for either density. Accordingly, any density combination that fulfills the condition above yields a Saha equilibrium. Similarly, pressure variations do not influence the equilibrium. To define a specific density and pressure, it is therefore necessary to apply integrated versions of the Boltzmann equation when solving the model. (In contrast to this, the temperature is T_h is unambiguous, or $< f >_{cIV} |_{z \to \infty}$ would not be zero.) Such integrated versions of the Boltzmann equation will therefore replace some of the differential equations later on.

Now that the boundary conditions have been identified, the conservation laws for mass, momentum, and energy will be looked at next. They can be derived as moments of the Boltzmann equation and ensure the physical validity of the model.

4.3.2. Conservation of Mass

The first equation is the conservation of mass, also called the continuity equation. It is given by adding the equations of the zeroth moment (weighting function m_s) for all species s.

$$\sum_s \nabla \cdot \mathbf{\Gamma}_s = \sum_s \int m_s < f_s >_c \mathrm{d}^3 v \qquad (4.35)$$

$\mathbf{\Gamma}_s = m_s n_s \mathbf{u}_s$ is the particle mass flux of species s with the corresponding particle mass m_s, density n_s, and average velocity $\mathbf{u}_s$. Since electrons are considered as massless $(m_e \approx 0, m_i \approx m_a)$, conservation of mass requires that

$$\nabla \cdot \mathbf{\Gamma}_{\mathrm{tot}} = \nabla \cdot (\mathbf{\Gamma}_i + \mathbf{\Gamma}_a) \,.$$

This leads to the relation

$$\nabla \cdot \mathbf{\Gamma}_{\mathrm{tot}} = \int m_i \left(< f_i >_c + < f_a >_c \right) \mathrm{d}^3 v \,.$$

To investigate this further, the composition of the collision terms is recalled:

$$< f_i >_c + < f_a >_c = < f_i >_{cI} + < f_i >_{cII} + < f_i >_{cIII} + < f_i >_{cIV} \\ + < f_a >_{cI} + < f_a >_{cII} + < f_a >_{cIII} + < f_a >_{cIV}$$

4. Construction of a Kinetic Model

Since $< f_i >_{cI} = - < f_a >_{cI}$ and $< f_i >_{cII} = - < f_a >_{cII}$, only the four terms $< f_i >_{cIII}$, $< f_a >_{cIII}$, $< f_i >_{cIV}$, and $< f_a >_{cIV}$ have to be considered. However, $< f_i >_{cIII}$ and $< f_a >_{cIII}$ are Boltzmann collision terms, which have the conservation laws included in their definition. Thus, only $< f_i >_{cIV}$ and $< f_a >_{cIV}$ remain:

$$
\begin{aligned}
\nabla \cdot \Gamma_{\text{tot}} &= \int_V m_i \left(< f_i >_c + < f_a >_c \right) \mathrm{d}^3 v \\
&= m_i \nu_1 \int_V \Bigg(3\big(f_i(z,\boldsymbol{v}) + f_a(z,\boldsymbol{v})\big) + \boldsymbol{v} \cdot \nabla_{\boldsymbol{v}}\big(f_i(z,\boldsymbol{v}) + f_a(z,\boldsymbol{v})\big) \\
&\qquad\qquad + \frac{T_h}{m_i}\nabla_{\boldsymbol{v}}^2\big(f_i(z,\boldsymbol{v}) + f_a(z,\boldsymbol{v})\big) \Bigg) \mathrm{d}^3 v \\
&= m_i \nu_1 \int_V \Bigg(3\big(f_i(z,\boldsymbol{v}) + f_a(z,\boldsymbol{v})\big) + \nabla_{\boldsymbol{v}} \cdot \big(\boldsymbol{v}\big(f_i(z,\boldsymbol{v}) + f_a(z,\boldsymbol{v})\big)\big) \\
&\qquad\qquad - 3\big(f_i(z,\boldsymbol{v}) + f_a(z,\boldsymbol{v})\big) + \frac{T_h}{m_i}\nabla_{\boldsymbol{v}}^2\big(f_i(z,\boldsymbol{v}) + f_a(z,\boldsymbol{v})\big) \Bigg) \mathrm{d}^3 v \\
&= m_i \nu_1 \int_V \nabla_{\boldsymbol{v}} \cdot \Bigg(\boldsymbol{v}\big(f_i(z,\boldsymbol{v}) + f_a(z,\boldsymbol{v})\big) + \frac{T_h}{m_i}\nabla_{\boldsymbol{v}}\big(f_i(z,\boldsymbol{v}) + f_a(z,\boldsymbol{v})\big) \Bigg) \mathrm{d}^3 v \\
&= m_i \nu_1 \int_{\partial V} \Bigg(\boldsymbol{v}\big(f_i(z,\boldsymbol{v}) + f_a(z,\boldsymbol{v})\big) + \frac{T_h}{m_i}\nabla_{\boldsymbol{v}}\big(f_i(z,\boldsymbol{v}) + f_a(z,\boldsymbol{v})\big) \Bigg) \mathrm{d}^2 \boldsymbol{f} \\
&= 0
\end{aligned}
$$

$$(4.36)$$

In the notation above, V is the volume of the complete v-space, with the corresponding volume element $\mathrm{d}v$. Its surface at infinity is ∂V with the vectorial surface element $\mathrm{d}^2 \boldsymbol{f}$ pointing outwards. Since $f_i = 0$ at infinity and since it can be assumed that f_i tends to zero faster than any power of v^{-n} as $v \to \infty$, the end result is zero.

The result can also be written in integrated form, representing the flux balance. It was assumed earlier that there is only one spatial dimension, z. Accordingly, the divergence becomes the derivative with respect to z and the integration has to be performed over z only. With the integration constant equal to zero (since there is no net particle gain or loss), one obtains an algebraic relation between the z-components of the ion and neutral fluxes:

$$
\Gamma_{iz} + \Gamma_{az} = 0 \qquad\qquad (4.37)
$$

4.3.3. Conservation of Momentum

The second equation is the conservation of momentum. It is given similarly to the equation for mass conservation, but with weight function $m_s\boldsymbol{v}$:

$$\sum_s \nabla \cdot \underline{\underline{\boldsymbol{\Pi}}}_s = \sum_s \left(q_s n_s \boldsymbol{E} + \int_V m_s \boldsymbol{v} < f_s >_c \mathrm{d}^3 v \right) \tag{4.38}$$

$\underline{\underline{\boldsymbol{\Pi}}}_s = m_s n_s \boldsymbol{u}_s \boldsymbol{u}_s + \underline{\underline{\boldsymbol{P}}}_s$ represents the momentum density flux of species s with the corresponding particle charge q_s and pressure tensor $\underline{\underline{\boldsymbol{P}}}_s = T_s n_s \underline{\underline{\boldsymbol{I}}}$. Accordingly,

$$\nabla \cdot \underline{\underline{\boldsymbol{\Pi}}}_{\mathrm{tot}} = \nabla \cdot \left(\underline{\underline{\boldsymbol{\Pi}}}_i + \underline{\underline{\boldsymbol{\Pi}}}_a \right) = e n_i \boldsymbol{E} + \int_V m_i \boldsymbol{v} \left(< f_i >_c + < f_a >_c \right) \mathrm{d}^3 v.$$

Following the reasoning applied in the mass conservation section, this can be simplified further:

$$
\begin{aligned}
\nabla \cdot \underline{\underline{\boldsymbol{\Pi}}}_{\mathrm{tot}} &= e n_i \boldsymbol{E} + \int_V m_i \boldsymbol{v} \left(< f_i >_c + < f_a >_c \right) \mathrm{d}^3 v \\
&= e n_i \boldsymbol{E} + \int_V m_i \boldsymbol{v} \left(< f_i >_{cIV} + < f_a >_{cIV} \right) \mathrm{d}^3 v \\
&= e n_i \boldsymbol{E} + m_i \nu_1 \int_V \boldsymbol{v} \Big(3\big(f_i(z,\boldsymbol{v}) + f_a(z,\boldsymbol{v})\big) + \boldsymbol{v} \cdot \nabla_{\boldsymbol{v}}\big(f_i(z,\boldsymbol{v}) + f_a(z,\boldsymbol{v})\big) \\
&\qquad + \frac{T_h}{m_i} \nabla_{\boldsymbol{v}}^2 \big(f_i(z,\boldsymbol{v}) + f_a(z,\boldsymbol{v})\big) \Big) \mathrm{d}^3 v \\
&= e n_i \boldsymbol{E} + m_i \nu_1 \int_V \boldsymbol{v} \nabla_{\boldsymbol{v}} \cdot \Big(\boldsymbol{v}\big(f_i(z,\boldsymbol{v}) + f_a(z,\boldsymbol{v})\big) + \frac{T_h}{m_i}\nabla_{\boldsymbol{v}}\big(f_i(z,\boldsymbol{v}) + f_a(z,\boldsymbol{v})\big) \Big) \mathrm{d}^3 v \\
&= e n_i \boldsymbol{E} - m_i \nu_1 \int_V \left(\boldsymbol{v}\big(f_i(z,\boldsymbol{v}) + f_a(z,\boldsymbol{v})\big) + \frac{T_h}{m_i}\nabla_{\boldsymbol{v}}\big(f_i(z,\boldsymbol{v}) + f_a(z,\boldsymbol{v})\big) \right) \mathrm{d}^3 v \\
&= e n_i \boldsymbol{E} - \nu_1 \left(\boldsymbol{\Gamma}_i + \boldsymbol{\Gamma}_a \right) = -T_e \frac{\mathrm{d}n_i}{\mathrm{d}z} \boldsymbol{e}_\parallel
\end{aligned}
\tag{4.39}
$$

For the last step, the flux balance equation and the self-consistent definition of the electric field were substituted. Integration of this equation is strongly simplified because of the one-dimensional approach in space. Taking the total pressure $p = n_{ip} T_h (1 + \gamma + \beta)$ (see equation (3.7)) as integration constant, one finds the pressure balance, which yields an algebraic relation between n_i and n_a:

$$m_i(n_i u_{i\parallel}^2 + n_a u_{a\parallel}^2) + T_i n_i + T_a n_a = -T_e n_i + p \qquad (4.40)$$

In the equation above, T_i and T_a are not constant but calculated from the distribution functions f_i and f_a, respectively.

4.3.4. Conservation of Energy

The third equation is the conservation of energy. It is given by the second (contracted) moment, with weight function $m_s v^2/2$:

$$\sum_s \nabla \cdot \Gamma_s^e = \sum_s \left(q_s n_s u_s \cdot E + \int_V \frac{m_s v^2}{2} < f_s >_c \mathrm{d}^3 v \right) \qquad (4.41)$$

$\Gamma_s^e = u_s(m_s n_s u_s^2 + 3 n_s T_s)/2 + q_s + \underline{\underline{P}}_s \cdot u_s$ is the energy flux of species s with the corresponding kinetic temperature T_s. Conservation of energy requires that

$$\nabla \cdot \Gamma_{\text{tot}}^e = \nabla \cdot (\Gamma_i^e + \Gamma_a^e) = e n_i u_i \cdot E + \int_V \frac{m_i v^2}{2} (< f_i >_c + < f_a >_c) \mathrm{d}^3 v.$$

Again, the arguments presented in the mass conservation discussion are applied, which yields a new relation to ensure energy conservation:

$$\nabla \cdot \Gamma^e_{tot} = e n_i \boldsymbol{u}_i \cdot \boldsymbol{E} + \int_V \frac{m_i \boldsymbol{v}^2}{2} \left(< f_i >_c + < f_a >_c \right) \mathrm{d}^3 v$$

$$= e n_i \boldsymbol{u}_i \cdot \boldsymbol{E} + \int_V \frac{m_i \boldsymbol{v}^2}{2} \left(< f_i >_{cIV} + < f_a >_{cIV} \right) \mathrm{d}^3 v$$

$$= e n_i \boldsymbol{u}_i \cdot \boldsymbol{E} + \frac{m_i \nu_1}{2} \int_V \boldsymbol{v}^2 \Big(3 \big(f_i(z, \boldsymbol{v}) + f_a(z, \boldsymbol{v}) \big) + \boldsymbol{v} \cdot \nabla_{\boldsymbol{v}} \big(f_i(z, \boldsymbol{v}) + f_a(z, \boldsymbol{v}) \big)$$
$$+ \frac{T_h}{m_i} \nabla^2_{\boldsymbol{v}} \big(f_i(z, \boldsymbol{v}) + f_a(z, \boldsymbol{v}) \big) \Big) \mathrm{d}^3 v$$

$$= e n_i \boldsymbol{u}_i \cdot \boldsymbol{E} + \frac{m_i \nu_1}{2} \int_V \boldsymbol{v}^2 \nabla_{\boldsymbol{v}} \cdot \Big(\boldsymbol{v} \big(f_i(z, \boldsymbol{v}) + f_a(z, \boldsymbol{v}) \big)$$
$$+ \frac{T_h}{m_i} \nabla_{\boldsymbol{v}} \big(f_i(z, \boldsymbol{v}) + f_a(z, \boldsymbol{v}) \big) \Big) \mathrm{d}^3 v$$

$$= e n_i \boldsymbol{u}_i \cdot \boldsymbol{E} + \frac{m_i \nu_1}{2} \Big[\int_V \boldsymbol{v}^2 \nabla_{\boldsymbol{v}} \cdot \big(\boldsymbol{v} \big(f_i(z, \boldsymbol{v}) + f_a(z, \boldsymbol{v}) \big) \big) \mathrm{d}^3 v$$
$$+ \int_V \boldsymbol{v}^2 \nabla_{\boldsymbol{v}} \cdot \Big(\frac{T_h}{m_i} \nabla_{\boldsymbol{v}} \big(f_i(z, \boldsymbol{v}) + f_a(z, \boldsymbol{v}) \big) \Big) \mathrm{d}^3 v \Big]$$

$$= e n_i \boldsymbol{u}_i \cdot \boldsymbol{E} + \frac{m_i \nu_1}{2} \Big[-2 \int_V \boldsymbol{v}^2 \big(f_i(z, \boldsymbol{v}) + f_a(z, \boldsymbol{v}) \big) \mathrm{d}^3 v$$
$$+ 6 \frac{T_h}{m_i} \int_V \big(f_i(z, \boldsymbol{v}) + f_a(z, \boldsymbol{v}) \big) \mathrm{d}^3 v \Big]$$

$$= e n_i \boldsymbol{u}_i \cdot \boldsymbol{E} + \frac{m_i \nu_1}{2} \Big[6 \frac{T_h}{m_i} (n_i + n_a) - 2 \int_V \boldsymbol{v}^2 \big(f_i(z, \boldsymbol{v}) + f_a(z, \boldsymbol{v}) \big) \mathrm{d}^3 v \Big]$$

$$= e n_i \boldsymbol{u}_i \cdot \boldsymbol{E} + 3 \nu_1 T_h (n_i + n_a) - 2 \nu_1 (e_i + e_a)$$

4.4. Mathematical Formulation of the Model

4.4.1. Normalization

The use of non-normalized variables has been helpful so far to clearly show the connection between physics and equations. However, the next sections involve some rather complex calculations. By normalizing all variables, the equations become more abstract and concise, which helps to focus on the mathematical details.

Additionally, the normalization introduces certain smallness parameters. These parameters represent physical scales within the system and allow certain physically consistent simplifications.

4. Construction of a Kinetic Model

Unless otherwise stated, all variables will be normalized from here. It is therefore not necessary to use different symbols to differentiate between normalized and non-normalized variables. The normalization used is based on the normalization introduced for the fluid model (see chapter 3). Since the temperature $T_{i/a}$ is not assumed to be constant anymore, it also needs to be normalized, which is done with respect to the temperature in the Saha plasma, T_h:

$$z \to \lambda_{ia}(1+\gamma)^{-1}z \quad n \to n_{tp}n \qquad f \to n_{tp}v_{as}^{-3}f$$

$$\boldsymbol{v} \to v_{as}\boldsymbol{v} \qquad \boldsymbol{E} \to T_h(\lambda_{ia}e)^{-1}\boldsymbol{E} \quad T \to T_h T$$

$$n_{tp} = n_{ip} + n_{ap} \qquad v_{as} = \sqrt{T_h/m_i} \qquad \lambda_{ia} = v_{as}(n_{ip}k_i)^{-1}$$

$$\alpha_{ii} = W_{ii}n_{tp}/k_i \qquad \alpha_{ai} = W_{ai}n_{tp}/k_i \qquad \alpha_{ia} = W_{ia}n_{tp}/k_i$$

$$\alpha_{aa} = W_{aa}n_{tp}/k_i \qquad \beta = T_e/T_h \qquad \gamma = n_{ap}/n_{ip} = n_{ip}k_r/k_i$$

$$\mu = m_e/m_i \qquad \tilde{\nu}_1 = (n_{tp}k_i)^{-1}\nu_1$$

With these rules, one obtains the normalized collision terms

$$< f_i >_{cI} = n_i(z)f_a(z, \boldsymbol{v}), \tag{4.42}$$

$$< f_i >_{cII} = -\gamma(1+\gamma)n_i(z)^2 f_i(z, \boldsymbol{v}), \tag{4.43}$$

$$< f_i >_{cIII} = \int_{\Omega'} \int_{\Omega} \int_{g'} g'^3 \left[\alpha_{ii}\left(f_i\left(\boldsymbol{v} - \frac{g'\boldsymbol{e}}{2} + \frac{g'\boldsymbol{e}'}{2}\right) f_i\left(\boldsymbol{v} - \frac{g'\boldsymbol{e}}{2} - \frac{g'\boldsymbol{e}'}{2}\right) \right.\right.$$

$$- f_i(\boldsymbol{v})f_i(\boldsymbol{v} - g'\boldsymbol{e}')\Big)$$

$$+ \alpha_{ia}\left(f_i\left(\boldsymbol{v} - \frac{g'\boldsymbol{e}}{2} + \frac{g'\boldsymbol{e}'}{2}\right) f_a\left(\boldsymbol{v} - \frac{g'\boldsymbol{e}}{2} - \frac{g'\boldsymbol{e}'}{2}\right) \right.$$

$$\left.\left. - f_i(\boldsymbol{v})f_a(\boldsymbol{v} - g'\boldsymbol{e}')\right)\right] \tag{4.44}$$

$$\times \mathrm{d}g'\mathrm{d}\Omega\mathrm{d}\Omega',$$

$$< f_i >_{cIV} = \tilde{\nu}_1\left(3f_i + \boldsymbol{v} \cdot \nabla_{\boldsymbol{v}}f_i + \nabla_{\boldsymbol{v}}^2 f_i\right), \tag{4.45}$$

$$< f_a >_{cI} = - n_i(z) f_a(z, \boldsymbol{v}), \tag{4.46}$$

$$< f_a >_{cII} = \gamma(1+\gamma) n_i(z)^2 f_i(z, \boldsymbol{v}), \tag{4.47}$$

$$
\begin{aligned}
< f_a >_{cIII} = \int_{\Omega'} \int_{\Omega} \int_{g'} g'^3 \bigg[& \alpha_{aa} \bigg(f_a \bigg(\boldsymbol{v} - \frac{g'\boldsymbol{e}}{2} + \frac{g'\boldsymbol{e}'}{2} \bigg) f_a \bigg(\boldsymbol{v} - \frac{g'\boldsymbol{e}}{2} - \frac{g'\boldsymbol{e}'}{2} \bigg) \\
& - f_a(\boldsymbol{v}) f_a(\boldsymbol{v} - g'\boldsymbol{e}') \bigg) \\
& + \alpha_{ai} \bigg(f_a \bigg(\boldsymbol{v} - \frac{g'\boldsymbol{e}}{2} + \frac{g'\boldsymbol{e}'}{2} \bigg) f_i \bigg(\boldsymbol{v} - \frac{g'\boldsymbol{e}}{2} - \frac{g'\boldsymbol{e}'}{2} \bigg) \\
& - f_a(\boldsymbol{v}) f_i(\boldsymbol{v} - g'\boldsymbol{e}') \bigg) \bigg] \\
& \times \mathrm{d}g' \mathrm{d}\Omega \mathrm{d}\Omega',
\end{aligned}
\tag{4.48}
$$

$$< f_a >_{cIV} = \tilde{\nu}_1 \left(3 f_a + \boldsymbol{v} \cdot \nabla_v f_a + \nabla_v^2 f_a \right), \tag{4.49}$$

and eventually the complete kinetic equations (4.27) and (4.33) in normalized form:

$$
\begin{aligned}
v_{\parallel} \frac{\partial f_i}{\partial z} &- \frac{\beta}{n_i} \frac{\mathrm{d}n_i}{\mathrm{d}z} \frac{\partial f_i}{\partial v_{\parallel}} \\
= & \, n_i f_a - \gamma(1+\gamma) n_i^2 f_i + \tilde{\nu}_1 \left(3 f_i + \boldsymbol{v} \cdot \nabla_v f_i + \nabla_v^2 f_i \right) \\
& + \int_{\Omega'} \int_{\Omega} \int_{g'} g'^3 \bigg[\alpha_{ii} \bigg(f_i \bigg(\boldsymbol{v} - \frac{g'\boldsymbol{e}}{2} + \frac{g'\boldsymbol{e}'}{2} \bigg) f_i \bigg(\boldsymbol{v} - \frac{g'\boldsymbol{e}}{2} - \frac{g'\boldsymbol{e}'}{2} \bigg) - f_i(\boldsymbol{v}) f_i(\boldsymbol{v} - g'\boldsymbol{e}') \bigg) \\
& + \alpha_{ia} \bigg(f_i \bigg(\boldsymbol{v} - \frac{g'\boldsymbol{e}}{2} + \frac{g'\boldsymbol{e}'}{2} \bigg) f_a \bigg(\boldsymbol{v} - \frac{g'\boldsymbol{e}}{2} - \frac{g'\boldsymbol{e}'}{2} \bigg) - f_i(\boldsymbol{v}) f_a(\boldsymbol{v} - g'\boldsymbol{e}') \bigg) \bigg] \\
& \times \mathrm{d}g' \mathrm{d}\Omega \mathrm{d}\Omega'
\end{aligned}
\tag{4.50}
$$

$$
\begin{aligned}
v_{\parallel} \frac{\partial f_a}{\partial z} \\
= & -n_i f_a + \gamma(1+\gamma) n_i^2 f_i + \tilde{\nu}_1 \left(3 f_a + \boldsymbol{v} \cdot \nabla_v f_a + \nabla_v^2 f_a \right) \\
& + \int_{\Omega'} \int_{\Omega} \int_{g'} g'^3 \bigg[\alpha_{aa} \bigg(f_a \bigg(\boldsymbol{v} - \frac{g'\boldsymbol{e}}{2} + \frac{g'\boldsymbol{e}'}{2} \bigg) f_a \bigg(\boldsymbol{v} - \frac{g'\boldsymbol{e}}{2} - \frac{g'\boldsymbol{e}'}{2} \bigg) - f_a(\boldsymbol{v}) f_a(\boldsymbol{v} - g'\boldsymbol{e}') \bigg) \\
& + \alpha_{ai} \bigg(f_a \bigg(\boldsymbol{v} - \frac{g'\boldsymbol{e}}{2} + \frac{g'\boldsymbol{e}'}{2} \bigg) f_i \bigg(\boldsymbol{v} - \frac{g'\boldsymbol{e}}{2} - \frac{g'\boldsymbol{e}'}{2} \bigg) - f_a(\boldsymbol{v}) f_i(\boldsymbol{v} - g'\boldsymbol{e}') \bigg) \bigg] \\
& \times \mathrm{d}g' \mathrm{d}\Omega \mathrm{d}\Omega'
\end{aligned}
\tag{4.51}
$$

Equations (4.50) and (4.51) are non-linear partial differential equations (PDEs) of the second order (due to the ∇_v^2 in the right hand side of the ion equation). To simplify this problem, the distribution functions f_i and f_a are expanded in terms of Hermite and Laguerre polynomials. This expansion reduces the second order PDE system to a system of (still non-linear) first-order ordinary equations. The price for this simplification is that instead of two equations, a larger number of equations has to be solved. This number is determined by the order of the expansion and usually amounts to ten and above. The following section will introduce the expansion in detail.

4.4.2. Polynomial Expansion

To separate the dependence of the distribution functions f_i and f_a on z from the dependence on the velocity $\boldsymbol{v}$, f_i and f_a are written as a product of two functions, depending on z and $\boldsymbol{v}$ only:

$$f(z, v_\parallel, v_\perp) = \tilde{a}(z)g(\boldsymbol{v}) \tag{4.52}$$

The function $g(\boldsymbol{v}) = g(v_x, v_y, v_z)$ is then expanded in terms of modified Hermite polynomials of the following form (see also section A.3), where $\mathrm{H}_m(v)$ represents the common definition of the Hermite polynomials:

$$\mathrm{He}_m(v) = 2^{-\frac{m}{2}} \frac{\mathrm{H}_m(v/\sqrt{2})}{\sqrt{m!}}$$

The three-dimensional function $g(v_x, v_y, v_z)$ can now be expanded into such polynomials straightforwardly:

$$g(v_x, v_y, v_z) = \sum_{m=0}^{\infty}\sum_{k=0}^{\infty}\sum_{l=0}^{\infty} a_{m,k,l}\mathrm{He}_m(v_z)\mathrm{He}_k(v_x)\mathrm{He}_l(v_y)e^{-(v_x^2+v_y^2+v_z^2)/2} \tag{4.53}$$

Besides the Hermite polynomials, an exponential term $e^{-(v_x^2+v_y^2+v_z^2)/2}$ was included. This exponential term is equal to the kernel of the orthogonality criterion for the Hermite functions. Although not necessary for the expansion itself, it is better to include it in the definition rather than the orthogonality criterion. The coefficients $a_{m,k,l}$ of the expansion are calculated as

$$a_{m,k,l} = \int_{-\infty}^{\infty} \int_{-\infty}^{\infty} \int_{-\infty}^{\infty} g(v_x, v_y, v_z) \text{He}_m(v_z) \text{He}_k(v_x) \text{He}_l(v_y) e^{-(v_x^2+v_y^2+v_z^2)/2} \mathrm{d}v_x \mathrm{d}v_y \mathrm{d}v_z.$$

(4.54)

Since the function $g(v_x, v_y, v_z)$ is radially symmetric, it can be written as $g(v_\parallel, v_\perp^2)$, with $v_z = v_\parallel$ and $v_x^2 + v_y^2 = v_\perp^2$. As another consequence, all parts of the expansion (4.53) that are odd with respect to v_x or v_y have to vanish. This is the case for Hermite polynomials with odd indices, thus $a_{m,k,l} = 0$ whenever either k or l are odd. Also, v_x and v_y are interchangeable, thus $a_{m,k,l} = a_{m,l,k}$. Altogether, this leads to a new expansion:

$$g(v_\parallel, v_\perp^2) = \sum_{m=0}^{\infty} \sum_{k'=0}^{\infty} \sum_{l'=0}^{\infty} a_{m,k',l'} \text{He}_m(v_z) \text{He}_{2k'}(v_x) \text{He}_{2l'}(v_y) e^{-(v_x^2+v_y^2+v_z^2)/2}$$

(4.55)

The respective coefficients $a_{m,k',l'}$ are calculated as

$$a_{m,k',l'} = \int_{-\infty}^{\infty} \int_{-\infty}^{\infty} \int_{-\infty}^{\infty} g(v_\parallel, v_\perp^2) \text{He}_m(v_z) \text{He}_{2k'}(v_x) \text{He}_{2l'}(v_y) e^{-(v_x^2+v_y^2+v_z^2)/2} \mathrm{d}v_x \mathrm{d}v_y \mathrm{d}v_z.$$

(4.56)

The prime will be dropped from now on.

Rewriting equation (4.56) in terms of cylindric coordinates $(v_\perp, \varphi, v_\parallel)$ leads to a more compact form of the expansion:

$$\begin{aligned} a_{m,k,l} &= \int_{-\infty}^{\infty} \int_{-\infty}^{\infty} \int_{-\infty}^{\infty} g(v_\parallel, v_\perp^2) \text{He}_m(v_z) \text{He}_{2k}(v_x) \text{He}_{2l}(v_y) e^{-(v_x^2+v_y^2+v_z^2)/2} \mathrm{d}v_x \mathrm{d}v_y \mathrm{d}v_z \\ &= \int_{-\infty}^{\infty} \int_{0}^{2\pi} \int_{0}^{\infty} g(v_\parallel, v_\perp^2) \text{He}_m(v_\parallel) \\ &\quad \times \text{He}_{2k}(v_\perp \cos\varphi) \text{He}_{2l}(v_\perp \sin\varphi) e^{-(v_\perp^2+v_\parallel^2)/2} v_\perp \mathrm{d}v_\perp \mathrm{d}\varphi \mathrm{d}v_\parallel. \end{aligned}$$

(4.57)

The integration over φ is carried out first, only the terms dependent on φ, i.e. $\text{He}_{2k}(v_\perp \cos\varphi)$ and $\text{He}_{2l}(v_\perp \sin\varphi)$, have to be taken into account. The result transfers the two-fold Hermite expansion into a single expansion into Laguerre polynomials (definition (A.38)):

$$\int_{0}^{2\pi} \text{He}_{2k}(v_\perp \cos\varphi) \text{He}_{2l}(v_\perp \sin\varphi) \mathrm{d}\varphi = \text{La}_{k+l}(v_\perp^2) 2^{1-k-l} \pi \frac{\sqrt{(2k)!(2l)!}}{k!l!}$$

(4.58)

4. Construction of a Kinetic Model

This formula can be proven by applying the sum definition of the Hermite and Laguerre polynomials ((A.21) and (A.38)), reordering the coefficients of the resulting double sum and then applying Vandermonde's identity (A.59). Substituting equation (4.58) into equation (4.57) yields

$$a_{m,k,l} = \frac{2^{1-k-l}\pi\sqrt{(2k)!(2l)!}}{k!l!} \int_{-\infty}^{\infty} \int_{0}^{\infty} g(v_{\parallel}, v_{\perp}^2) \mathrm{He}_m(v_{\parallel}) \mathrm{La}_{k+l}(v_{\perp}^2) e^{-\left(v_{\perp}^2 + v_{\parallel}^2\right)/2} v_{\perp} dv_{\perp} dv_{\parallel},$$

$$(4.59)$$

which defines the more compact expansion and its coefficients $a_{m,n}$:

$$g(v_{\parallel}, v_{\perp}^2) = \sum_{m=0}^{\infty} \sum_{n=0}^{\infty} a_{m,n} \mathrm{He}_m(v_{\parallel}) \mathrm{La}_n(v_{\perp}^2) e^{-\left(v_{\parallel}^2 + v_{\perp}^2\right)/2} \tag{4.60}$$

$$a_{m,n} = \int_{-\infty}^{\infty} \int_{0}^{\infty} g(v_{\parallel}, v_{\perp}^2) \mathrm{He}_m(v_{\parallel}) \mathrm{La}_n(v_{\perp}^2) e^{-\left(v_{\perp}^2 + v_{\parallel}^2\right)/2} v_{\perp} dv_{\perp} dv_{\parallel} \tag{4.61}$$

Comparison of equations (4.59) and (4.61) yields the relation between the coefficients $a_{m,n}$ and $a_{m,k,l}$:

$$a_{m,k,l} = \frac{2^{1-k-l}\pi\sqrt{(2k)!(2l)!}}{k!l!} a_{m,n}, \quad \text{with} \quad n = k + l \tag{4.62}$$

For the actual model, the dependence on the spatial coordinate z has to be included, which is implemented by simply combining the coefficients $a_{m,k,l}$ with the function $\tilde{a}(z)$: $a_{m,k,l}(z) := \tilde{a}(z)a_{m,k,l}$. Also, it is necessary to limit the length of the sums in (4.53) to a finite number of elements, namely M, K, and L. This last restriction limits the accuracy of the expansion and as a general condition for the expansion to be valid it can be stated that all coefficients with indices $M + 1$, $K + 1$, $L + 1$, or higher have to be negligibly small.

In summary, the distribution functions for the heavy particles are written as follows, where an extra factor $1/\sqrt{2\pi}^3$ has been included for normalization reasons:

$$f_i(z, v_\parallel, v_\perp^2) = \sum_{m=0}^{M} \sum_{n=0}^{N} \frac{a_{m,n}(z)}{\sqrt{2\pi}^3} \mathrm{He}_m(v_\parallel) \mathrm{La}_n(v_\perp^2) e^{-\left(v_\parallel^2 + v_\perp^2\right)/2} \qquad (4.63)$$

$$f_a(z, v_\parallel, v_\perp^2) = \sum_{m=0}^{M} \sum_{n=0}^{N} \frac{b_{m,n}(z)}{\sqrt{2\pi}^3} \mathrm{He}_m(v_\parallel) \mathrm{La}_n(v_\perp^2) e^{-\left(v_\parallel^2 + v_\perp^2\right)/2} \qquad (4.64)$$

In reverse, the coefficients $a_{m,n}(z)$ and $b_{m,n}(z)$ for a given distribution function f_i or f_a can be calculated from:

$$a_{m,n}(z) = \int_{-\infty}^{\infty} \int_0^\infty \int_0^{2\pi} f_i(z, v_\parallel, v_\perp^2) \mathrm{He}_m(v_\parallel) \mathrm{La}_n(v_\perp^2) v_\perp \mathrm{d}\varphi \mathrm{d}v_\perp \mathrm{d}v_\parallel \qquad (4.65)$$

$$b_{m,n}(z) = \int_{-\infty}^{\infty} \int_0^\infty \int_0^{2\pi} f_a(z, v_\parallel, v_\perp^2) \mathrm{He}_m(v_\parallel) \mathrm{La}_n(v_\perp^2) v_\perp \mathrm{d}\varphi \mathrm{d}v_\perp \mathrm{d}v_\parallel \qquad (4.66)$$

Some important moments of the distribution function are given here in simple terms of the coefficients $a_{m,n}(z)$ and $b_{m,n}(z)$ for demonstrational purposes and further reference:

$$n_i(z) = \int f_i(z, \boldsymbol{v}) \mathrm{d}^3 v = \iiint f_i(z, v_\parallel, v_\perp^2) v_\perp \mathrm{d}v_\perp \mathrm{d}\varphi \mathrm{d}v_\parallel = a_{0,0}(z) \qquad (4.67)$$

$$n_a(z) = b_{0,0}(z) \qquad (4.68)$$

$$\Gamma_{iz}(z) = n_i(z) u_{iz} = \iiint \boldsymbol{v}_\parallel f_i(z, v_\parallel, v_\perp^2) v_\perp \mathrm{d}v_\perp \mathrm{d}\varphi \mathrm{d}v_\parallel = a_{1,0}(z) \qquad (4.69)$$

$$\Gamma_{az}(z) = n_a(z) u_{az} = b_{1,0}(z) \qquad (4.70)$$

$$n_i(z) \boldsymbol{u}_{i\parallel}^2(z) = \iiint v_\parallel^2 f_i(z, v_\parallel, v_\perp^2) v_\perp \mathrm{d}v_\perp \mathrm{d}\varphi \mathrm{d}v_\parallel - \iiint f_i(z, v_\parallel, v_\perp^2) v_\perp \mathrm{d}v_\perp \mathrm{d}\varphi \mathrm{d}v_\parallel$$

$$= \sqrt{2} a_{2,0}(z) \qquad (4.71)$$

$$n_a(z) \boldsymbol{u}_{a\parallel}^2(z) = \sqrt{2} b_{2,0}(z) \qquad (4.72)$$

4. Construction of a Kinetic Model

$$
n_i(z)\boldsymbol{u}_{i\perp}^2(z) = \iiint v_\perp^2 f_i(z, v_\parallel, v_\perp^2) v_\perp \mathrm{d}v_\perp \mathrm{d}\varphi \mathrm{d}v_\parallel - \iiint f_i(z, v_\parallel, v_\perp^2) v_\perp \mathrm{d}v_\perp \mathrm{d}\varphi \mathrm{d}v_\parallel
$$

$$
= a_{0,0}(z) + 2a_{0,1}(z) \tag{4.73}
$$

$$
n_a(z)\boldsymbol{u}_{a\perp}^2(z) = b_{0,0}(z) + 2b_{0,1}(z) \tag{4.74}
$$

$$
e_i(z) = \iiint \frac{v_\parallel^2 + v_\perp^2}{2} f_i(z, v_\parallel, v_\perp^2) v_\perp \mathrm{d}v_\perp \mathrm{d}\varphi \mathrm{d}v_\parallel
$$

$$
= \frac{3}{2}a_{0,0}(z) + \frac{1}{\sqrt{2}}a_{2,0}(z) + a_{0,1}(z) \tag{4.75}
$$

$$
e_a(z) = \frac{3}{2}b_{0,0}(z) + \frac{1}{\sqrt{2}}b_{2,0}(z) + b_{0,1}(z) \tag{4.76}
$$

$$
T_i(z) = \frac{1}{3n_i(z)} \iiint (\boldsymbol{v} - \boldsymbol{u}_i)^2 f_i(z, \boldsymbol{v}) v_\perp \mathrm{d}v_\perp \mathrm{d}\varphi \mathrm{d}v_\parallel = \frac{2e_i(z)}{3n_i(z)} - \frac{\boldsymbol{u}_{i\parallel}^2}{3}
$$

$$
= 1 + \frac{\sqrt{2}a_{2,0}(z)}{3a_{0,0}(z)} + \frac{2a_{0,1}(z)}{3a_{0,0}(z)} - \frac{a_{1,0}^2(z)}{3a_{0,0}^2(z)} \tag{4.77}
$$

$$
T_a(z) = \frac{2e_a(z)}{3n_a(z)} - \frac{\boldsymbol{u}_{a\parallel}^2}{3} = 1 + \frac{\sqrt{2}b_{2,0}(z)}{3b_{0,0}(z)} + \frac{2b_{0,1}(z)}{3b_{0,0}(z)} - \frac{b_{1,0}^2(z)}{3b_{0,0}^2(z)} \tag{4.78}
$$

5. Solution for the Kinetic Model

The solution to the model presented in the previous chapter consists of two steps. The first step is to substitute definitions (4.63) and (4.64) into the flux and pressure balances as well as into equation (4.1). By applying the orthogonality criteria for the Hermite and Laguerre polynomials, a set of algebraic and ordinary differential equations for the coefficients $a_{m,n}(z)$ and $b_{m,n}(z)$ is derived. The second step is to solve this equation set, thus to obtain the coefficients and heavy particle distribution functions f_i and f_a. Although the idea behind each step is rather straightforward, their realization turns out to be quite involved. During the first step, the elastic heavy particle collisions require special attention. For the second step, a high-dimensional boundary problem needs to be solved. Both steps are discussed in more detail below.

5.1. The Equation System

The coefficient version of the balance equations (4.37) and (4.40) can be derived rather simply by substituting the coefficient representations of the fluid dynamic quantities (4.67) - (4.78):

$$a_{1,0}(z) + b_{1,0}(z) = 0 \qquad (5.1)$$

$$\frac{4\sqrt{2}}{3}\left(a_{2,0}(z) + b_{2,0}(z)\right) + (1+\beta)a_{0,0}(z)$$
$$+ b_{0,0}(z) + \frac{2}{3}\left(\frac{a_{1,0}^2}{a_{0,0}} + \frac{b_{1,0}^2}{b_{0,0}}\right) = \frac{1+\beta+\gamma}{1+\gamma} \qquad (5.2)$$

The equations for the remaining coefficients are differential equations and derived from the Boltzmann equation (4.1), more specifically from its less general versions (4.25) and (4.33). Beginning with the left hand side of equation (4.50), one finds

5. Solution for the Kinetic Model

$$v_{\parallel}\frac{\partial f_i}{\partial z} - \frac{\beta}{n_i}\frac{\mathrm{d}n_i}{\mathrm{d}z}\frac{\partial f_i}{\partial v_{\parallel}}. \tag{5.3}$$

With the definition of f_i as given in (4.63) and the recurrence relations for the modified Hermite polynomials, it is possible to find new expressions for the terms $v_{\parallel}\partial f_i/\partial z$ and $\partial f_i/\partial v_{\parallel}$:

$$v_{\parallel}\frac{\partial f_i}{\partial z} = \sum_{m=0}^{M}\sum_{n=0}^{N}\left(\frac{\sqrt{m}}{\sqrt{2\pi}^3}\frac{\mathrm{d}a_{m-1,n}(z)}{\mathrm{d}z} + \frac{\sqrt{m+1}}{\sqrt{2\pi}^3}\frac{\mathrm{d}a_{m+1,n}(z)}{\mathrm{d}z}\right)\mathrm{He}_m(v_{\parallel})\mathrm{La}_n(v_{\perp}^2)e^{-(v_{\parallel}^2+v_{\perp}^2)/2} \tag{5.4}$$

$$\frac{\partial f_i}{\partial v_{\parallel}} = -\sum_{m=0}^{M}\sum_{n=0}^{N}\frac{\sqrt{m}}{\sqrt{2\pi}^3}a_{m-1,n}(z)\mathrm{He}_m(v_{\parallel})\mathrm{La}_n(v_{\perp}^2)e^{-(v_{\parallel}^2+v_{\perp}^2)/2} \tag{5.5}$$

Substituting these expressions into equation (4.50) then yields:

$$\frac{1}{\sqrt{2\pi}^3}\sum_{m=0}^{M}\sum_{n=0}^{N}\left(\sqrt{m}\frac{\mathrm{d}a_{m-1,n}(z)}{\mathrm{d}z} + \sqrt{m+1}\frac{\mathrm{d}a_{m+1,n}(z)}{\mathrm{d}z} + \frac{\beta}{a_{0,0}(z)}\frac{\mathrm{d}a_{0,0}(z)}{\mathrm{d}z}\sqrt{m}a_{m-1,n}(z)\right)$$
$$\times\,\mathrm{He}_m(v_{\parallel})\mathrm{La}_n(v_{\perp}^2)e^{-(v_{\parallel}^2+v_{\perp}^2)/2} = <f_i>_c$$

Integration on both sides according to the orthonormality criteria (A.22) and (A.40) yields a differential equation for the coefficients $a_{m,n}$:

$$\sqrt{m}\frac{\mathrm{d}a_{m-1,n}(z)}{\mathrm{d}z} + \sqrt{m+1}\frac{\mathrm{d}a_{m+1,n}(z)}{\mathrm{d}z} + \frac{\beta}{a_{0,0}(z)}\frac{\mathrm{d}a_{0,0}(z)}{\mathrm{d}z}\sqrt{m}a_{m-1,n}(z)$$
$$= 2\pi\int_{-\infty}^{\infty}\int_{0}^{\infty} <f_i>_c\,\mathrm{He}_m(v_{\parallel})\mathrm{La}_n(v_{\perp}^2)v_{\perp}\mathrm{d}v_{\perp}\mathrm{d}v_{\parallel} \tag{5.6}$$

The equation also contains coefficients of other indices ($m+1$ and $m-1$). Therefore, to solve the equation, it is necessary to combine its instances for different indices m.

The respective equation for the neutral particles is easily obtained by omitting the term caused by the electric field:

$$\sqrt{m}\frac{\mathrm{d}b_{m-1,n}(z)}{\mathrm{d}z} + \sqrt{m+1}\frac{\mathrm{d}b_{m+1,n}(z)}{\mathrm{d}z}$$
$$= 2\pi \int_{-\infty}^{\infty}\int_{0}^{\infty} < f_a >_c \mathrm{He}_m(v_{\parallel})\mathrm{La}_n(v_{\perp}^2)v_{\perp}\mathrm{d}v_{\perp}\mathrm{d}v_{\parallel} \tag{5.7}$$

To complete the equation system, it is now necessary to treat the collision terms on the right hand side accordingly. Due to the similarities of the expressions for the corresponding neutral and ion collision terms, it is sufficient to discuss the ion collision terms in length and transfer the results to the neutral collision terms.

5.1.1. Ionization and Recombination: $< f_{i/a} >_{cI}$ and $< f_{i/a} >_{cII}$

Combining the definition of the ionization term (4.4) with the definition of the neutral distribution function (4.64) and the orthogonality relations for the Hermite and Laguerre polynomials (eqs. (A.22) and (A.40)) yields:

$$2\pi \int_{-\infty}^{\infty}\int_{0}^{\infty} < f_i >_{cI} \mathrm{He}_m(v_{\parallel})\mathrm{La}_n(v_{\perp}^2)v_{\perp}\mathrm{d}v_{\perp}\mathrm{d}v_{\parallel}$$
$$= 2\pi \int_{-\infty}^{\infty}\int_{0}^{\infty} n_i(z)f_a(z,\boldsymbol{v})\mathrm{He}_m(v_{\parallel})\mathrm{La}_n(v_{\perp}^2)v_{\perp}\mathrm{d}v_{\perp}\mathrm{d}v_{\parallel}$$
$$= 2\pi \int_{-\infty}^{\infty}\int_{0}^{\infty} a_{0,0}(z)\sum_{\tilde{m}=0}^{\widetilde{M}}\sum_{\tilde{n}=0}^{\widetilde{N}}\frac{b_{\tilde{m},\tilde{n}}(z)}{\sqrt{2\pi}^3}\mathrm{He}_{\tilde{m}}(v_{\parallel})\mathrm{La}_{\tilde{n}}(v_{\perp}^2)e^{-(v_{\parallel}^2+v_{\perp}^2)/2}$$
$$\times \mathrm{He}_m(v_{\parallel})\mathrm{La}_n(v_{\perp}^2)v_{\perp}\mathrm{d}v_{\perp}\mathrm{d}v_{\parallel}$$
$$= a_{0,0}(z)\sum_{\tilde{m}=0}^{\widetilde{M}}\sum_{\tilde{n}=0}^{\widetilde{N}}\frac{b_{\tilde{m},\tilde{n}}(z)}{\sqrt{2\pi}}\int_{-\infty}^{\infty}\int_{0}^{\infty}\mathrm{He}_{\tilde{m}}(v_{\parallel})\mathrm{La}_{\tilde{n}}(v_{\perp}^2)$$
$$\times \mathrm{He}_m(v_{\parallel})\mathrm{La}_n(v_{\perp}^2)e^{-(v_{\parallel}^2+v_{\perp}^2)/2}v_{\perp}\mathrm{d}v_{\perp}\mathrm{d}v_{\parallel}$$
$$= a_{0,0}(z)b_{m,n}(z)$$

$$\tag{5.8}$$

Due to their similar structure, the recombination term for the ions and the ionization

5. Solution for the Kinetic Model

and recombination terms for the neutrals can be derived from this result directly:

$$2\pi \int_{-\infty}^{\infty} \int_{0}^{\infty} < f_i >_{cII} \mathrm{He}_m(v_\parallel) \mathrm{La}_n(v_\perp^2) v_\perp \mathrm{d}v_\perp \mathrm{d}v_\parallel = -\gamma(1+\gamma)a_{0,0}(z)^2 a_{m,n}(z) \quad (5.9)$$

$$2\pi \int_{-\infty}^{\infty} \int_{0}^{\infty} < f_a >_{cI} \mathrm{He}_m(v_\parallel) \mathrm{La}_n(v_\perp^2) v_\perp \mathrm{d}v_\perp \mathrm{d}v_\parallel = -a_{0,0}(z) b_{m,n}(z) \quad (5.10)$$

$$2\pi \int_{-\infty}^{\infty} \int_{0}^{\infty} < f_a >_{cII} \mathrm{He}_m(v_\parallel) \mathrm{La}_n(v_\perp^2) v_\perp \mathrm{d}v_\perp \mathrm{d}v_\parallel = \gamma(1+\gamma)a_{0,0}(z)^2 a_{m,n}(z) \quad (5.11)$$

5.1.2. Elastic Collisions: $< f_{i/a} >_{cIII}$

By applying the orthonomality relation to equation (4.23), one finds the collision term for elastic collisions in basic form:

$$
\begin{aligned}
&2\pi \int_{-\infty}^{\infty} \int_{0}^{\infty} < f_i >_{cIII} \mathrm{He}_m(v_\parallel) \mathrm{La}_n(v_\perp^2) v_\perp \mathrm{d}v_\perp \mathrm{d}v_\parallel \\
&= 2\pi \int_{-\infty}^{\infty} \int_{0}^{\infty} \\
&\times \int_{\Omega'} \int_{\Omega} \int_{g'} g'^3 \left[\alpha_{ii}\left(f_i\left(v - \frac{g'e}{2} + \frac{g'e'}{2}\right) f_i\left(v - \frac{g'e}{2} - \frac{g'e'}{2}\right) - f_i(v)f_i(v - g'e)\right) \right. \\
&\qquad\qquad \left. + \alpha_{ia}\left(f_i\left(v - \frac{g'e}{2} + \frac{g'e'}{2}\right) f_a\left(v - \frac{g'e}{2} - \frac{g'e'}{2}\right) - f_i(v)f_a(v - g'e)\right)\right] \\
&\qquad \times \mathrm{d}g' \mathrm{d}\Omega \mathrm{d}\Omega' \\
&\qquad \times \mathrm{He}_m(v_\parallel) \mathrm{La}_n(v_\perp^2) v_\perp \mathrm{d}v_\perp \mathrm{d}v_\parallel
\end{aligned}
$$

$$(5.12)$$

To use this term for further calculations, it is necessary to align the arguments of the distribution functions. To this aim, v is shifted to $v + g'e/2$ in equation (5.12). It is important to move the Hermite and Laguerre functions under the inner integral before doing so, since they become functions of g' with this shift. The integration limits for v on the other hand do not need to be changed, since they entail all possible velocities:

$$2\pi \int_{-\infty}^{\infty} \int_{0}^{\infty} < f_i >_{cIII} \mathrm{He}_m(v_{\parallel})\mathrm{La}_n(v_{\perp}^2)v_{\perp}dv_{\perp}dv_{\parallel}$$

$$= 2\pi \int_{-\infty}^{\infty} \int_{0}^{\infty} \int_{\Omega'} \int_{\Omega} \int_{g'} g'^3$$

$$\times \left[\alpha_{ii} \left[\mathrm{He}_m\left(\left(\boldsymbol{v} + \frac{g'\boldsymbol{e}}{2} \right)_{\parallel} \right) \mathrm{La}_n\left(\left(\boldsymbol{v} + \frac{g'\boldsymbol{e}}{2} \right)_{\perp}^2 \right) f_i\left(\boldsymbol{v} + \frac{g'\boldsymbol{e}'}{2} \right) f_i\left(\boldsymbol{v} - \frac{g'\boldsymbol{e}'}{2} \right) \right. \right.$$

$$\left. - \mathrm{He}_m\left(\left(\boldsymbol{v} + \frac{g'\boldsymbol{e}}{2} \right)_{\parallel} \right) \mathrm{La}_n\left(\left(\boldsymbol{v} + \frac{g'\boldsymbol{e}}{2} \right)_{\perp}^2 \right) f_i\left(\boldsymbol{v} + \frac{g'\boldsymbol{e}}{2} \right) f_i\left(\boldsymbol{v} - \frac{g'\boldsymbol{e}}{2} \right) \right]$$

$$+ \alpha_{ia} \left[\mathrm{He}_m\left(\left(\boldsymbol{v} + \frac{g'\boldsymbol{e}}{2} \right)_{\parallel} \right) \mathrm{La}_n\left(\left(\boldsymbol{v} + \frac{g'\boldsymbol{e}}{2} \right)_{\perp}^2 \right) f_i\left(\boldsymbol{v} + \frac{g'\boldsymbol{e}'}{2} \right) f_a\left(\boldsymbol{v} - \frac{g'\boldsymbol{e}'}{2} \right) \right.$$

$$\left. \left. - \mathrm{He}_m\left(\left(\boldsymbol{v} + \frac{g'\boldsymbol{e}}{2} \right)_{\parallel} \right) \mathrm{La}_n\left(\left(\boldsymbol{v} + \frac{g'\boldsymbol{e}}{2} \right)_{\perp}^2 \right) f_i\left(\boldsymbol{v} + \frac{g'\boldsymbol{e}}{2} \right) f_a\left(\boldsymbol{v} - \frac{g'\boldsymbol{e}}{2} \right) \right] \right]$$

$$\times \, dg'd\Omega d\Omega'$$
$$\times \, v_{\perp}dv_{\perp}dv_{\parallel}$$

$$(5.13)$$

The vectors $\boldsymbol{e}$ and $\boldsymbol{e}'$ can be formally interchanged, because of the similar integrals for $d\Omega$ and $d\Omega'$. Doing so for the negative terms under the integral leads to a more concise form:

$$2\pi \int_{-\infty}^{\infty} \int_{0}^{\infty} < f_i >_{cIII} \mathrm{He}_m(v_{\parallel})\mathrm{La}_n(v_{\perp}^2)v_{\perp}dv_{\perp}dv_{\parallel}$$

$$= 2\pi \int_{-\infty}^{\infty} \int_{0}^{\infty} \int_{\Omega'} \int_{\Omega} \int_{g'} g'^3$$

$$\times \left[\alpha_{ii} f_i\left(\boldsymbol{v} + \frac{g'\boldsymbol{e}'}{2} \right) f_i\left(\boldsymbol{v} - \frac{g'\boldsymbol{e}'}{2} \right) + \alpha_{ia} f_i\left(\boldsymbol{v} + \frac{g'\boldsymbol{e}'}{2} \right) f_a\left(\boldsymbol{v} - \frac{g'\boldsymbol{e}'}{2} \right) \right]$$

$$\times \left[\mathrm{He}_m\left(\left(\boldsymbol{v} + \frac{g'\boldsymbol{e}'}{2} \right)_{\parallel} \right) \mathrm{La}_n\left(\left(\boldsymbol{v} + \frac{g'\boldsymbol{e}'}{2} \right)_{\perp}^2 \right) \right.$$

$$(5.14)$$

$$\left. - \mathrm{He}_m\left(\left(\boldsymbol{v} + \frac{g'\boldsymbol{e}'}{2} \right)_{\parallel} \right) \mathrm{La}_n\left(\left(\boldsymbol{v} + \frac{g'\boldsymbol{e}'}{2} \right)_{\perp}^2 \right) \right]$$

$$\times \, dg'd\Omega d\Omega'$$
$$\times \, v_{\perp}dv_{\perp}dv_{\parallel}$$

The treatment of this integral needs to be investigated in more detail. For better readability, this will be done for the ion-atom part (α_{ia}) only. The ion-ion (α_{ii}) part can later be added by formally adjusting the indices and adding it to the ion-atom part again.

The distribution functions f_i and f_a are written in terms of Hermite and Laguerre polynomials, as defined in equations (4.63) and (4.64). Expressions for the parallel and perpendicular parts of $v \pm g'e/2$ and $v \pm g'e'/2$ are needed. Whereas the parallel parts are simply given by

$$\left(v \pm \frac{g'e}{2} \right)_{\parallel} = v_{\parallel} \pm \frac{g' \cos \theta}{2}, \tag{5.15}$$

$$\left(v \pm \frac{g'e'}{2} \right)_{\parallel} = v_{\parallel} \pm \frac{g' \cos \theta'}{2}, \tag{5.16}$$

the perpendicular parts need to be calculated following the law of cosines:

$$\left(v \pm \frac{g'e}{2} \right)_{\perp}^{2} = v_{\perp}^2 + \frac{g'^2 \sin^2 \theta}{4} \pm v_{\perp} g' \sin \theta \cos(\varphi_0 - \varphi) \tag{5.17}$$

$$\left(v \pm \frac{g'e'}{2} \right)_{\perp}^{2} = v_{\perp}^2 + \frac{g'^2 \sin^2 \theta'}{4} \pm v_{\perp} g' \sin \theta' \cos(\varphi_0 - \varphi') \tag{5.18}$$

Here, φ_0 is the initial direction of $v_{\perp}$. It is only given for completeness and can be omitted in the following calculations, because the integration over φ always covers the whole interval $[0; 2\pi)$. Substituting equations (4.63) and (4.64) for f_i and f_a now yields a complete expression for the elastic collisions between ions and neutrals:

$$2\pi \int_{-\infty}^{\infty} \int_{0}^{\infty} < f_i >_{cIIIia} \mathrm{He}_m(v_\parallel)\mathrm{La}_n(v_\perp^2)v_\perp \mathrm{d}v_\perp \mathrm{d}v_\parallel$$

$$= 2\pi \int_{-\infty}^{\infty} \int_{0}^{\infty} \int_{\Omega'} \int_{\Omega} \int_{g'} \alpha_{ia} g'^3$$

$$\times \left[\mathrm{He}_m\left(v_\parallel + \frac{g'\cos\theta}{2}\right)\mathrm{La}_n\left(v_\perp^2 + \frac{g'^2\sin^2\theta}{4} + v_\perp g'\sin\theta\cos\varphi\right)\right.$$

$$\left. - \mathrm{He}_m\left(v_\parallel + \frac{g'\cos\theta'}{2}\right)\mathrm{La}_n\left(v_\perp^2 + \frac{g'^2\sin^2\theta'}{4} + v_\perp g'\sin\theta'\cos\varphi'\right)\right]$$

$$\times \sum_{\widetilde{m}=0}^{\widetilde{M}} \sum_{\widetilde{n}=0}^{\widetilde{N}} \left[\frac{a_{\widetilde{m},\widetilde{n}}(z)}{\sqrt{2\pi}^3} \mathrm{He}_{\widetilde{m}}\left(v_\parallel + \frac{g'\cos\theta'}{2}\right)\mathrm{La}_{\widetilde{n}}\left(v_\perp^2 + \frac{g'^2\sin^2\theta'}{4} + v_\perp g'\sin\theta'\cos\varphi'\right)\right.$$

$$\left. \times e^{-(\boldsymbol{v}+\boldsymbol{g}'\boldsymbol{e}'/2)^2/2}\right]$$

$$\times \sum_{\widetilde{\widetilde{m}}=0}^{\widetilde{\widetilde{M}}} \sum_{\widetilde{\widetilde{n}}=0}^{\widetilde{\widetilde{N}}} \left[\frac{b_{\widetilde{\widetilde{m}},\widetilde{\widetilde{n}}}(z)}{\sqrt{2\pi}^3} \mathrm{He}_{\widetilde{\widetilde{m}}}\left(v_\parallel - \frac{g'\cos\theta'}{2}\right)\mathrm{La}_{\widetilde{\widetilde{n}}}\left(v_\perp^2 + \frac{g'^2\sin^2\theta'}{4} - v_\perp g'\sin\theta'\cos\varphi'\right)\right.$$

$$\left. \times e^{-(\boldsymbol{v}-\boldsymbol{g}'\boldsymbol{e}'/2)^2/2}\right]\mathrm{d}g'\mathrm{d}\Omega\mathrm{d}\Omega'v_\perp \mathrm{d}v_\perp \mathrm{d}v_\parallel$$

$$= \sum_{\widetilde{m}=0}^{\widetilde{M}} \sum_{\widetilde{n}=0}^{\widetilde{N}} \sum_{\widetilde{\widetilde{m}}=0}^{\widetilde{\widetilde{M}}} \sum_{\widetilde{\widetilde{n}}=0}^{\widetilde{\widetilde{N}}} \frac{a_{\widetilde{m},\widetilde{n}}(z)b_{\widetilde{\widetilde{m}},\widetilde{\widetilde{n}}}(z)}{4\pi^2} \int_{-\infty}^{\infty} \int_{0}^{\infty} \int_{\Omega'} \int_{\Omega} \int_{g'} \alpha_{ia} g'^3$$

$$\times \left[\mathrm{He}_m\left(v_\parallel + \frac{g'\cos\theta}{2}\right)\mathrm{La}_n\left(v_\perp^2 + \frac{g'^2\sin^2\theta}{4} + v_\perp g'\sin\theta\cos\varphi\right)\right.$$

$$\left. - \mathrm{He}_m\left(v_\parallel + \frac{g'\cos\theta'}{2}\right)\mathrm{La}_n\left(v_\perp^2 + \frac{g'^2\sin^2\theta'}{4} + v_\perp g'\sin\theta'\cos\varphi'\right)\right]$$

$$\times \mathrm{He}_{\widetilde{m}}\left(v_\parallel + \frac{g'\cos\theta'}{2}\right)\mathrm{La}_{\widetilde{n}}\left(v_\perp^2 + \frac{g'^2\sin^2\theta'}{4} + v_\perp g'\sin\theta'\cos\varphi'\right)$$

$$\times \mathrm{He}_{\widetilde{\widetilde{m}}}\left(v_\parallel - \frac{g'\cos\theta'}{2}\right)\mathrm{La}_{\widetilde{\widetilde{n}}}\left(v_\perp^2 + \frac{g'^2\sin^2\theta'}{4} - v_\perp g'\sin\theta'\cos\varphi'\right)$$

$$\times e^{-(v_\parallel^2+v_\perp^2+g'^2/4)}\mathrm{d}g'\mathrm{d}\Omega\mathrm{d}\Omega'v_\perp \mathrm{d}v_\perp \mathrm{d}v_\parallel$$

$$=: \sum_{\widetilde{m}=0}^{\widetilde{M}} \sum_{\widetilde{n}=0}^{\widetilde{N}} \sum_{\widetilde{\widetilde{m}}=0}^{\widetilde{\widetilde{M}}} \sum_{\widetilde{\widetilde{n}}=0}^{\widetilde{\widetilde{N}}} \alpha_{ia} \frac{a_{\widetilde{m},\widetilde{n}}(z)b_{\widetilde{\widetilde{m}},\widetilde{\widetilde{n}}}(z)}{4\pi^2} \left(A_{m,\widetilde{m},\widetilde{\widetilde{m}},n,\widetilde{n},\widetilde{\widetilde{n}}} - B_{m,\widetilde{m},\widetilde{\widetilde{m}},n,\widetilde{n},\widetilde{\widetilde{n}}}\right)$$

$$= \sum_{\widetilde{m}=0}^{\widetilde{M}} \sum_{\widetilde{n}=0}^{\widetilde{N}} \sum_{\widetilde{\widetilde{m}}=0}^{\widetilde{\widetilde{M}}} \sum_{\widetilde{\widetilde{n}}=0}^{\widetilde{\widetilde{N}}} \alpha_{ia} \frac{a_{\widetilde{m},\widetilde{n}}(z)b_{\widetilde{\widetilde{m}},\widetilde{\widetilde{n}}}(z)}{4\pi^2} C_{m,\widetilde{m},\widetilde{\widetilde{m}},n,\widetilde{n},\widetilde{\widetilde{n}}}$$

$$(5.19)$$

The combination of all elements $A_{m,\tilde{m},\tilde{\tilde{m}},n,\tilde{n},\tilde{\tilde{n}}}$, $B_{m,\tilde{m},\tilde{\tilde{m}},n,\tilde{n},\tilde{\tilde{n}}}$, and $C_{m,\tilde{m},\tilde{\tilde{m}},n,\tilde{n},\tilde{\tilde{n}}}$ can be written as three matrices $\underline{\underline{A}}$, $\underline{\underline{B}}$, and $\underline{\underline{C}}$. In principle, all that needs to be done now is to calculate the matrix elements by solving the integrals and then to evaluate the sums. However, in reality this turns out to be a rather difficult task. The upper limit for the indices $m, \tilde{m}, \tilde{\tilde{m}}, n, \tilde{n}$, and $\tilde{\tilde{n}}$ is defined by the upper limit of the polynomial expansion. For example, given an upper limit of 10 for each expansion requires multiplication of several polynomials of the order 10. This by itself takes some time already. Additionally, results have to be obtained for all possible combinations, which requires calculation and storage of 10^6 elements for the given example. Fortunately, many of the elements are zero, these elements do not need to be calculated. The calculation of the other elements is further simplified by symmetries within the matrices. Despite all that, the amount of necessary calculations remains large and finding a reasonable limit for the expansions is crucial to avoid extremely long calculations that lead to no substantial gain in accuracy.

Calculation of the Matrix Elements

The purpose of this section is to investigate the calculation of the collision term (5.19). To keep terms as concise as possible, expressions that are constant with respect to the integrals are omitted. They simply can be multiplied with the final result. Also, the sums are omitted at this point and all considerations are made for one summand. Furthermore, it is helpful to split the term (5.19) and treat $A_{m,\tilde{m},\tilde{\tilde{m}},n,\tilde{n},\tilde{\tilde{n}}}$ and $B_{m,\tilde{m},\tilde{\tilde{m}},n,\tilde{n},\tilde{\tilde{n}}}$ separately. Starting with $A_{m,\tilde{m},\tilde{\tilde{m}},n,\tilde{n},\tilde{\tilde{n}}}$, one finds:

$$
\begin{aligned}
A_{m,\tilde{m},\tilde{\tilde{m}},n,\tilde{n},\tilde{\tilde{n}}} = \int_{-\infty}^{\infty} & \int_{0}^{\infty} \int_{\Omega'} \int_{\Omega} \int_{g'} g'^3 \\
\times \mathrm{He}_m & \left(v_{\parallel} + \frac{g' \cos\theta}{2} \right) \mathrm{La}_n \left(v_{\perp}^2 + \frac{g'^2 \sin^2\theta}{4} + v_{\perp} g' \sin\theta \cos\varphi \right) \\
\times \mathrm{He}_{\tilde{m}} & \left(v_{\parallel} + \frac{g' \cos\theta'}{2} \right) \mathrm{La}_{\tilde{n}} \left(v_{\perp}^2 + \frac{g'^2 \sin^2\theta'}{4} + v_{\perp} g' \sin\theta' \cos\varphi' \right) \\
\times \mathrm{He}_{\tilde{\tilde{m}}} & \left(v_{\parallel} - \frac{g' \cos\theta'}{2} \right) \mathrm{La}_{\tilde{\tilde{n}}} \left(v_{\perp}^2 + \frac{g'^2 \sin^2\theta'}{4} - v_{\perp} g' \sin\theta' \cos\varphi' \right) \\
\times \, e^{-(v_{\parallel}^2 + v_{\perp}^2 + g'^2/4)} & \mathrm{d}g' \mathrm{d}\Omega \mathrm{d}\Omega' v_{\perp} \mathrm{d}v_{\perp} \mathrm{d}v_{\parallel}
\end{aligned}
$$

$$(5.20)$$

This term contains expressions that depend on either $v_{\perp}$ or $v_{\parallel}$ only. These expressions are separated accordingly, together with the corresponding integrals:

$$A_{m,\bar{m},\tilde{\tilde{m}},n,\bar{n},\tilde{\tilde{n}}} = \int_{\Omega'} \int_{\Omega} \int_{g'} g'^3 \left[\int_{-\infty}^{\infty} \mathrm{He}_m \left(v_{\parallel} + \frac{g'\cos\theta}{2} \right) \mathrm{He}_{\bar{m}} \left((v_{\parallel} + \frac{g'\cos\theta'}{2} \right) \right.$$
$$\times \mathrm{He}_{\tilde{\tilde{m}}} \left((v_{\parallel} - \frac{g'\cos\theta'}{2} \right) e^{-v_{\parallel}^2} \mathrm{d}v_{\parallel} \Big]$$
$$\times \left[\int_0^{\infty} \mathrm{La}_n \left(v_{\perp}^2 + \frac{g'^2\sin^2\theta}{4} + v_{\perp}g'\sin\theta\cos\varphi \right) \right.$$
$$\times \mathrm{La}_{\bar{n}} \left(v_{\perp}^2 + \frac{g'^2\sin^2\theta'}{4} + v_{\perp}g'\sin\theta'\cos\varphi' \right)$$
$$\times \mathrm{La}_{\tilde{\tilde{n}}} \left(v_{\perp}^2 + \frac{g'^2\sin^2\theta'}{4} - v_{\perp}g'\sin\theta'\cos\varphi' \right)$$
$$\times v_{\perp} e^{-v_{\perp}^2} \mathrm{d}v_{\perp} \Big] e^{-g'^2/4} \mathrm{d}g'\mathrm{d}\Omega\mathrm{d}\Omega' \tag{5.21}$$

Obviously only the Laguerre polynomials contain φ and φ', so that the corresponding integrals can be moved:

$$A_{m,\bar{m},\tilde{\tilde{m}},n,\bar{n},\tilde{\tilde{n}}} = \int_0^{\pi} \int_0^{\pi} \int_0^{\infty} g'^3 \left[\int_{-\infty}^{\infty} \mathrm{He}_m \left(v_{\parallel} + \frac{g'\cos\theta}{2} \right) \mathrm{He}_{\bar{m}} \left(v_{\parallel} + \frac{g'\cos\theta'}{2} \right) \right.$$
$$\times \mathrm{He}_{\tilde{\tilde{m}}} \left(v_{\parallel} - \frac{g'\cos\theta'}{2} \right) e^{-v_{\parallel}^2} \mathrm{d}v_{\parallel} \Big]$$
$$\times \left[\int_0^{\infty} \int_0^{2\pi} \mathrm{La}_n \left(v_{\perp}^2 + \frac{g'^2\sin^2\theta}{4} + v_{\perp}g'\sin\theta\cos\varphi \right) \mathrm{d}\varphi \right.$$
$$\times \int_0^{2\pi} \mathrm{La}_{\bar{n}} \left(v_{\perp}^2 + \frac{g'^2\sin^2\theta'}{4} + v_{\perp}g'\sin\theta'\cos\varphi' \right)$$
$$\times \mathrm{La}_{\tilde{\tilde{n}}} \left(v_{\perp}^2 + \frac{g'^2\sin^2\theta'}{4} - v_{\perp}g'\sin\theta'\cos\varphi' \right) \mathrm{d}\varphi'$$
$$\times v_{\perp} e^{-v_{\perp}^2} \mathrm{d}v_{\perp} \Big] e^{-g'^2/4} \sin\theta\sin\theta'\mathrm{d}g'\mathrm{d}\theta\mathrm{d}\theta'$$
$$=: \int_0^{\pi} \int_0^{\pi} \int_0^{\infty} g'^3 I1(g',\theta,\theta')_{m,\bar{m},\tilde{\tilde{m}}} I2(g',\theta,\theta')_{n,\bar{n},\tilde{\tilde{n}}} e^{-g'^2/4} \sin\theta\sin\theta'\mathrm{d}g'\mathrm{d}\theta\mathrm{d}\theta' \tag{5.22}$$

By applying equation (A.29), the integral $I1(g',\theta,\theta')_{m,\bar{m},\tilde{\tilde{m}}}$ can be solved analytically:

5. Solution for the Kinetic Model

$$I1(g',\theta,\theta')_{m,\tilde{m},\overset{\approx}{m}} = \int_{-\infty}^{\infty} \mathrm{He}_m\left(v_{\parallel}+\frac{g'\cos\theta}{2}\right)$$

$$\times \mathrm{He}_{\tilde{m}}\left(v_{\parallel}+\frac{g'\cos\theta'}{2}\right)\mathrm{He}_{\overset{\approx}{m}}\left(v_{\parallel}-\frac{g'\cos\theta'}{2}\right)e^{-v_{\parallel}^2}dv_{\parallel}$$

$$= \frac{2^{-(m+\tilde{m}+\overset{\approx}{m})}}{\sqrt{m!\tilde{m}!\overset{\approx}{m}!}}\sum_{k_0=0}^{m}\sum_{k_1=0}^{\tilde{m}}\sum_{k_2=0}^{\overset{\approx}{m}}\binom{m}{k_0}\binom{\tilde{m}}{k_1}\binom{\overset{\approx}{m}}{k_2}$$

$$\times \mathrm{H}_{m-k_0}\left(\frac{g'\cos\theta}{2}\right)\mathrm{H}_{\tilde{m}-k_1}\left(\frac{g'\cos\theta'}{2}\right)\mathrm{H}_{\overset{\approx}{m}-k_2}\left(-\frac{g'\cos\theta'}{2}\right)$$

$$\times \begin{cases} \dfrac{\sqrt{\pi}2^{(k_0+k_1+k_2)/2}k_0!k_1!k_2!}{2} & \text{for} & \left.\begin{matrix} k_0+k_1+k_2 \quad \text{even} \\ \text{and} \\ k_0,k_1,k_2\le(k_0+k_1+k_2)/2 \end{matrix}\right\} \\ \prod_{i=0}^{2}((k_0+k_1+k_2)/2-k_i)! & & \\ 0 & \text{else} \end{cases}$$

$$(5.23)$$

By looking at the unevaluated integral it is easy to see that values of the integral $I1_{\overset{\approx}{m},\tilde{m},m}$ can be obtained by replacing "$\cos\theta'$" with "$-\cos\theta'$" in the results for $I1_{m,\tilde{m},\overset{\approx}{m}}$. After integrating over θ', this will become completely symmetric, so that

$$A_{m,\tilde{m},\overset{\approx}{m},n,\tilde{n},\overset{\approx}{n}} = A_{m,\overset{\approx}{m},\tilde{m},n,\tilde{n},\overset{\approx}{n}}.$$

Also, $I2_{n,\tilde{n},\overset{\approx}{n}} = I2_{n,\overset{\approx}{n},\tilde{n}}$, so that the matrix $\underline{\underline{A}}$ is symmetric with respect to $\tilde{n}$ and $\overset{\approx}{n}$ as well:

$$A_{m,n,\tilde{m},\overset{\approx}{m},\tilde{n},\overset{\approx}{n}} = A_{m,n,\tilde{m},\overset{\approx}{m},\overset{\approx}{n},\tilde{n}}.$$

Regarding the four-dimensional submatrix of $\underline{\underline{A}}$ for a fixed set of $\{m,n\}$, this means that only one eighth of the coefficients have to be calculated and stored. Additionally, the two-dimensional submatrix of $\underline{\underline{A}}$ for a fixed set of $\{m,\tilde{m},\overset{\approx}{m}\}$ consists of zero elements for $m+\tilde{m}+\overset{\approx}{m}$ =odd.

Now, the $B_{m,\tilde{m},\overset{\approx}{m},n,\tilde{n},\overset{\approx}{n}}$ in equation (5.19) need to be analyzed similarly:

$$B_{m,\tilde{m},\tilde{\tilde{m}},n,\tilde{n},\tilde{\tilde{n}}} = \int_{-\infty}^{\infty} \int_0^{\infty} \int_{g'} \int_{\Omega} \int_{\Omega'} g'^3$$

$$\times \text{He}_m \left(v_\parallel + \frac{g' \cos\theta'}{2} \right) \text{La}_n \left(v_\perp^2 + \frac{g'^2 \sin^2\theta'}{4} + v_\perp g' \sin\theta' \cos\varphi' \right)$$

$$\times \text{He}_{\tilde{m}} \left(v_\parallel + \frac{g' \cos\theta'}{2} \right) \text{La}_{\tilde{n}} \left(v_\perp^2 + \frac{g'^2 \sin^2\theta'}{4} + v_\perp g' \sin\theta' \cos\varphi' \right)$$

$$\times \text{He}_{\tilde{\tilde{m}}} \left(v_\parallel - \frac{g' \cos\theta'}{2} \right) \text{La}_{\tilde{\tilde{n}}} \left(v_\perp^2 + \frac{g'^2 \sin^2\theta'}{4} - v_\perp g' \sin\theta' \cos\varphi' \right)$$

$$\times e^{-(v_\parallel^2 + v_\perp^2 + g'^2/4)} \mathrm{d}g' \mathrm{d}\Omega \mathrm{d}\Omega' v_\perp \mathrm{d}v_\perp \mathrm{d}v_\parallel$$

$$= 4\pi \int_{-\infty}^{\infty} \int_0^{\infty} \int_{g'} \int_{\Omega'} g'^3$$

$$\times \text{He}_m \left(v_\parallel + \frac{g' \cos\theta'}{2} \right) \text{La}_n \left(v_\perp^2 + \frac{g'^2 \sin^2\theta'}{4} + v_\perp g' \sin\theta' \cos\varphi' \right)$$

$$\times \text{He}_{\tilde{m}} \left(v_\parallel + \frac{g' \cos\theta'}{2} \right) \text{La}_{\tilde{n}} \left(v_\perp^2 + \frac{g'^2 \sin^2\theta'}{4} + v_\perp g' \sin\theta' \cos\varphi' \right)$$

$$\times \text{He}_{\tilde{\tilde{m}}} \left(v_\parallel - \frac{g' \cos\theta'}{2} \right) \text{La}_{\tilde{\tilde{n}}} \left(v_\perp^2 + \frac{g'^2 \sin^2\theta'}{4} - v_\perp g' \sin\theta' \cos\varphi' \right)$$

$$\times e^{-(v_\parallel^2 + v_\perp^2 + g'^2/4)} \mathrm{d}g' \mathrm{d}\Omega' v_\perp \mathrm{d}v_\perp \mathrm{d}v_\parallel$$

$$= 4\pi \int_0^{\pi} \int_0^{\infty} g'^3$$

$$\times \left[\int_{-\infty}^{\infty} \text{He}_m \left(v_\parallel + \frac{g' \cos\theta'}{2} \right) \text{He}_{\tilde{m}} \left(v_\parallel + \frac{g' \cos\theta'}{2} \right) \right.$$

$$\times \left. \text{He}_{\tilde{\tilde{m}}} \left(v_\parallel - \frac{g' \cos\theta'}{2} \right) e^{-v_\parallel^2} \mathrm{d}v_\parallel \right]$$

$$\times \left[\int_0^{2\pi} \int_0^{\infty} \text{La}_n \left(v_\perp^2 + \frac{g'^2 \sin^2\theta'}{4} + v_\perp g' \sin\theta' \cos\varphi' \right) \right.$$

$$\times \text{La}_{\tilde{n}} \left(v_\perp^2 + \frac{g'^2 \sin^2\theta'}{4} + v_\perp g' \sin\theta' \cos\varphi' \right)$$

$$\times \text{La}_{\tilde{\tilde{n}}} \left(v_\perp^2 + \frac{g'^2 \sin^2\theta'}{4} - v_\perp g' \sin\theta' \cos\varphi' \right)$$

$$\left. \times e^{-v_\perp^2} v_\perp \mathrm{d}v_\perp \mathrm{d}\varphi' \right] e^{-g'^2/4} \sin\theta' \mathrm{d}g' \mathrm{d}\theta'$$

$$=: 4\pi \int_0^{\pi} \int_0^{\infty} g'^3 J1(g',\theta')_{m,\tilde{m},\tilde{\tilde{m}}} J2(g',\theta')_{n,\tilde{n},\tilde{\tilde{n}}} e^{-g'^2/4} \sin\theta' \mathrm{d}g' \mathrm{d}\theta'$$

$$(5.24)$$

$J1$ can be calculated analytically just like $I1$. The integration order over φ' and $v_\perp$

within $J2$ can be interchanged, integrating over φ' first proved to be faster. The matrix $\underline{\underline{B}}$ is obviously symmetric with respect to $\{m, \widetilde{m}\}$ and $\{n, \widetilde{n}\}$. With $\underline{\underline{A}}$ and $\underline{\underline{B}}$ known, the complete term including the ion-ion part can be calculated as

$$
2\pi \int_{-\infty}^{\infty} \int_{0}^{\infty} < f_i >_{cIII} \mathrm{He}_m(v_\parallel)\mathrm{La}_n(v_\perp^2)v_\perp \mathrm{d}v_\perp \mathrm{d}v_\parallel
$$
$$
= \sum_{\widetilde{m}=0}^{\widetilde{M}} \sum_{\widetilde{n}=0}^{\widetilde{N}} \sum_{\widetilde{\widetilde{m}}=0}^{\widetilde{\widetilde{M}}} \sum_{\widetilde{\widetilde{n}}=0}^{\widetilde{\widetilde{N}}} \frac{\alpha_{ii}a_{\widetilde{m},\widetilde{n}}(z)a_{\widetilde{\widetilde{m}},\widetilde{\widetilde{n}}}(z) + \alpha_{ia}a_{\widetilde{m},\widetilde{n}}(z)b_{\widetilde{\widetilde{m}},\widetilde{\widetilde{n}}}(z)}{4\pi^2} C_{m,n,\widetilde{m},\widetilde{\widetilde{m}},\widetilde{n},\widetilde{\widetilde{n}}}.
\tag{5.25}
$$

The corresponding neutral collision term is essentially the same:

$$
2\pi \int_{-\infty}^{\infty} \int_{0}^{\infty} < f_a >_{cIII} \mathrm{He}_m(v_\parallel)\mathrm{La}_n(v_\perp^2)v_\perp \mathrm{d}v_\perp \mathrm{d}v_\parallel
$$
$$
= \sum_{\widetilde{m}=0}^{\widetilde{M}} \sum_{\widetilde{n}=0}^{\widetilde{N}} \sum_{\widetilde{\widetilde{m}}=0}^{\widetilde{\widetilde{M}}} \sum_{\widetilde{\widetilde{n}}=0}^{\widetilde{\widetilde{N}}} \frac{\alpha_{aa}b_{\widetilde{m},\widetilde{n}}(z)b_{\widetilde{\widetilde{m}},\widetilde{\widetilde{n}}}(z) + \alpha_{ai}b_{\widetilde{m},\widetilde{n}}(z)a_{\widetilde{\widetilde{m}},\widetilde{\widetilde{n}}}(z)}{4\pi^2} C_{m,n,\widetilde{m},\widetilde{\widetilde{m}},\widetilde{n},\widetilde{\widetilde{n}}}.
\tag{5.26}
$$

Although $\underline{\underline{A}}$ and $\underline{\underline{B}}$ bear multiple symmetries and submatrices containing zeros only, the number of matrix elements to be calculated remains very high. Additionally, the integrals $I1, I2, J1$, and $J2$ contain high-order polynomials. Their computation takes up most of the total computing time. Using Wolfram Mathematica, sum limits of $M = \widetilde{M} = \widetilde{\widetilde{M}} = N = \widetilde{N} = \widetilde{\widetilde{N}} = 10$ ($\hat{=}11^6$ matrix elements) could not be realized due to insufficient memory. However, the number of elements decreases rapidly with lower limits. Since the expansion in $v_\perp$ is less critical than the expansion in $v_\parallel$, a lower limit for N is chosen to allow higher values of M. An upper limit of $M = \widetilde{M} = \widetilde{\widetilde{M}} = 7$ and $N = \widetilde{N} = \widetilde{\widetilde{N}} = 1$ seems to promise reasonable accuracy with feasible demands on memory and computation time. This combination is therefore chosen for the rest of this work.

5.1.3. Heat Sink Collision Terms: $< f_{i/a} >_{cIV}$

The collision terms for the last kind of collisions in the coefficient formulation can be calculated straightforwardly from definitions (4.24) and (4.32):

$$2\pi \int_{-\infty}^{\infty} \int_0^{\infty} <f_i>_{cIV} \mathrm{He}_m(v_{\parallel})\mathrm{La}_n(v_{\perp}^2)v_{\perp}\mathrm{d}v_{\perp}\mathrm{d}v_{\parallel}$$

$$= 2\pi \int_{-\infty}^{\infty} \int_0^{\infty} \tilde{\nu}_1 \Bigg(3 \sum_{m=0}^{M}\sum_{n=0}^{N} \frac{a_{m,n}(z)}{\sqrt{2\pi}^3}\mathrm{He}_m(v_{\parallel})\mathrm{La}_n(v_{\perp}^2)e^{-(v_{\parallel}^2+v_{\perp}^2)/2}$$

$$+\, \boldsymbol{v}\cdot\nabla_{\boldsymbol{v}} \sum_{m=0}^{M}\sum_{n=0}^{N} \frac{a_{m,n}(z)}{\sqrt{2\pi}^3}\mathrm{He}_m(v_{\parallel})\mathrm{La}_n(v_{\perp}^2)e^{-(v_{\parallel}^2+v_{\perp}^2)/2}$$

$$+\, \nabla_{\boldsymbol{v}}^2 \sum_{m=0}^{M}\sum_{n=0}^{N} \frac{a_{m,n}(z)}{\sqrt{2\pi}^3}\mathrm{He}_m(v_{\parallel})\mathrm{La}_n(v_{\perp}^2)e^{-(v_{\parallel}^2+v_{\perp}^2)/2} \Bigg)$$

$$\times\, \mathrm{He}_m(v_{\parallel})\mathrm{La}_n(v_{\perp}^2)v_{\perp}\mathrm{d}v_{\perp}\mathrm{d}v_{\parallel} \tag{5.27}$$

$$= -\tilde{\nu}_1(2n+m)a_{m,n}(z)$$

For the last step, the recurrence relations given in the appendix (A.3 and A.4) were used. Analogously, one finds:

$$2\pi \int_{-\infty}^{\infty} \int_0^{\infty} <f_a>_{cIV} \mathrm{He}_m(v_{\parallel})\mathrm{La}_n(v_{\perp}^2)v_{\perp}\mathrm{d}v_{\perp}\mathrm{d}v_{\parallel} = -\tilde{\nu}_1(2n+m)b_{m,n}(z) \tag{5.28}$$

5.1.4. All Terms Combined

Equations (5.6) and (5.7) can now be written explicitly:

$$\sqrt{m}\frac{\mathrm{d}a_{m-1,n}(z)}{\mathrm{d}z} + \sqrt{m+1}\frac{\mathrm{d}a_{m+1,n}(z)}{\mathrm{d}z} + \frac{\beta}{a_{0,0}(z)}\frac{\mathrm{d}a_{0,0}(z)}{\mathrm{d}z}\sqrt{m}a_{m-1,n}(z)$$

$$= a_{0,0}(z)b_{m,n}(z) - \gamma(1+\gamma)a_{0,0}(z)^2 a_{m,n}(z) - \tilde{\nu}_1(2n+m)a_{m,n}(z)$$

$$+ \sum_{\tilde{m}=0}^{\widetilde{M}}\sum_{\tilde{n}=0}^{\widetilde{N}}\sum_{\approx m=0}^{\approx M}\sum_{\approx n=0}^{\approx N} \frac{\alpha_{ii}a_{\tilde{m},\tilde{n}}(z)a_{\approx m,\approx n}(z) + \alpha_{ia}a_{\tilde{m},\tilde{n}}(z)b_{\approx m,\approx n}(z)}{4\pi^2}C_{m,n,\tilde{m},\tilde{n},\approx m,\approx n} \tag{5.29}$$

$$
\sqrt{m}\frac{\mathrm{d}b_{m-1,n}(z)}{\mathrm{d}z} + \sqrt{m+1}\frac{\mathrm{d}b_{m+1,n}(z)}{\mathrm{d}z}
$$
$$
= -a_{0,0}(z)b_{m,n}(z) + \gamma(1+\gamma)a_{0,0}(z)^2 a_{m,n}(z)
$$
$$
+ \sum_{\tilde{m}=0}^{\tilde{M}}\sum_{\tilde{n}=0}^{\tilde{N}}\sum_{\tilde{\tilde{m}}=0}^{\tilde{\tilde{M}}}\sum_{\tilde{\tilde{n}}=0}^{\tilde{\tilde{N}}} \frac{\alpha_{aa}b_{\tilde{m},\tilde{n}}(z)b_{\tilde{\tilde{m}},\tilde{\tilde{n}}}(z) + \alpha_{ai}b_{\tilde{m},\tilde{n}}(z)a_{\tilde{\tilde{m}},\tilde{\tilde{n}}}(z)}{4\pi^2}C_{m,n,\tilde{m},\tilde{\tilde{m}},\tilde{n},\tilde{\tilde{n}}}
$$

(5.30)

By permutation of all m and n between 0 and M or N, respectively, these expression yield $2(M+1)(N+1)$ equations to be used for the solution of the model. However, the equations contain $2(M+2)(N+1)$ variables, because for $m = M$, the equations contain $a_{M+1,n}$ and $b_{M+1,n}$. This can be rectified by simply omitting the extra terms, which is equivalent to a cutoff of the polynomial expansion at $m = M$ (see chapter 4.4.2). Now, these equations have to be completed by a set of boundary conditions to ensure physical validity. To recall the physical boundary conditions here, they are (i) a Saha plasma on the right and (ii) a space charge sheath on the left. The following section addresses the mathematical implementation of these boundary conditions.

5.2. Boundary Conditions

5.2.1. Saha Plasma at $z \to \infty$

Like in the fluid dynamic case, the final solution of the kinetic model will be carried out numerically. This also means that here, a condition given at infinity poses problems and needs to be transformed into an appropriate condition at a finite distance from the cathode. Again, this can be performed by a linearization around the Saha equilibrium. The equilibrium in kinetic terms is defined by a Maxwellian distribution function for f_i and f_a, and the equations to be linearized are the Boltzmann equations for the ions and atoms, respectively. By applying the orthogonal expansion presented in the previous chapter, this linearization is transferred into a linearization for the coefficients $a_{m,n}(z)$ and $b_{m,n}(z)$ around their corresponding equilibrium values $a_{m,n}(z \to \infty) = \bar{a}_{m,n}$ and $b_{m,n}(z \to \infty) = \bar{b}_{m,n}$:

$$
a_{m,n}(z) = \bar{a}_{m,n} + \delta a_{m,n}(z) \tag{5.31}
$$
$$
b_{m,n}(z) = \bar{b}_{m,n} + \delta b_{m,n}(z) \tag{5.32}
$$

By simply substituting a Maxwellian distribution for the distribution functions in (4.63) and (4.64), it becomes obvious that these equilibrium values are $\bar{a}_{0,0} = 1/(1+\gamma)$, $\bar{b}_{0,0} = \gamma/(1+\gamma)$, and $\bar{a}_{m,n} = \bar{b}_{m,n} = 0$ whenever m or n are larger than zero. The linearized equations then read:

$$
\begin{aligned}
\frac{\mathrm{d}}{\mathrm{d}z} & \left(\beta(\gamma+1)\sqrt{m}\bar{a}_{m-1,n}\delta a_{0,0}(z) + \sqrt{m}\delta a_{m-1,n}(z) + \sqrt{m+1}\delta a_{m+1,n}(z) \right) \\
&= \frac{1}{(\gamma+1)}\delta b_{m,n}(z) - \left(3\gamma\bar{a}_{m,n} - 2\bar{b}_{m,n} \right)\delta a_{0,0}(z) - \frac{\gamma+(m+2n)\tilde{\nu}_1}{\gamma+1}\delta a_{m,n}(z) \\
&\quad + \sum_{\tilde{m}=0}^{\widetilde{M}}\sum_{\tilde{n}=0}^{\widetilde{N}}\left(\frac{\alpha_{ii}(C_{m,\tilde{m},0,n,\tilde{n},0} + C_{m,0,\tilde{m},n,0,\tilde{n}}) + \gamma\alpha_{ia}C_{m,\tilde{m},0,n,\tilde{n},0}}{4\pi^2(\gamma+1)}\delta a_{\tilde{m},\tilde{n}}(z) \right. \\
&\qquad \left. + \frac{\alpha_{ia}C_{m,0,\tilde{m},n,0,\tilde{n}}}{4\pi^2(\gamma+1)}\delta b_{\tilde{m},\tilde{n}}(z) \right)
\end{aligned}
\tag{5.33}
$$

$$
\begin{aligned}
\frac{\mathrm{d}}{\mathrm{d}z} & \left(\sqrt{m}\delta b_{m-1,n}(z) + \sqrt{m+1}\delta b_{m+1,n}(z) \right) \\
&= \frac{\gamma}{\gamma+1}\delta a_{m,n}(z) - \left(\bar{b}_{m,n} - 2\gamma\bar{a}_{m,n} \right)\delta a_{0,0}(z) - \frac{1+(m+2n)\tilde{\nu}_1}{\gamma+1}\delta b_{m,n}(z) \\
&\quad + \sum_{\tilde{m}=0}^{\widetilde{M}}\sum_{\tilde{n}=0}^{\widetilde{N}}\left(\frac{\alpha_{aa}\gamma(C_{m,\tilde{m},0,n,\tilde{n},0} + C_{m,0,\tilde{m},n,0,\tilde{n}}) + \alpha_{ia}C_{m,\tilde{m},0,n,\tilde{n},0}}{4\pi^2(\gamma+1)}\delta b_{\tilde{m},\tilde{n}}(z) \right. \\
&\qquad \left. + \frac{\alpha_{ia}\gamma C_{m,0,\tilde{m},n,0,\tilde{n}}}{4\pi^2(\gamma+1)}\delta a_{\tilde{m},\tilde{n}}(z) \right)
\end{aligned}
\tag{5.34}
$$

Again, permutation of all m and n yields a total of $2(M+1)(N+1)$ equations. They form a system of homogeneous ordinary differential equations and can be written as a matrix equation of the following form:

$$
\underline{\underline{L}} \cdot \frac{\mathrm{d}\boldsymbol{f}}{\mathrm{d}z} = \underline{\underline{M}} \cdot \boldsymbol{f}
\tag{5.35}
$$

Here, $\boldsymbol{f} = (\delta a_{0,0}, \delta a_{1,0}, \dots, \delta a_{M-1,N}, \delta a_{M,N}, \delta b_{0,0}, \delta b_{1,0}, \dots, \delta b_{M-1,N}, \delta b_{M,N})^T$ is the vector that contains all the unknown functions $\delta a_{m,n}(z)$ and $\delta b_{m,n}(z)$. $\underline{\underline{L}}$ and $\underline{\underline{M}}$ are square matrices of the following (symbolic) forms:

$$\underline{\underline{M}} =
\begin{pmatrix}
\begin{aligned}
&\sum_{\bar{m}=0}^{\widetilde{M}}\sum_{\bar{n}=0}^{\widetilde{N}}\frac{\alpha_{ii}(C_{m,\bar{m},0,n,\bar{n},0}+C_{m,0,\bar{m},n,0,\bar{n}})+\gamma\alpha_{ia}C_{m,\bar{m},0,n,\bar{n},0}}{4\pi^2(\gamma+1)}\delta a_{\bar{m},\bar{n}}(z)\\
&\quad-(3\gamma\bar{a}_{m,n}-2\bar{b}_{m,n})\delta a_{0,0}(z)-\frac{\gamma+(m+2n)\tilde{\nu}_1}{\gamma+1}\delta a_{m,n}(z)
\end{aligned}
&
\sum_{\bar{m}=0}^{\widetilde{M}}\sum_{\bar{n}=0}^{\widetilde{N}}\frac{\alpha_{ia}C_{m,0,\bar{m},n,0,\bar{n}}}{4\pi^2(\gamma+1)}\delta b_{\bar{m},\bar{n}}(z)+\frac{1}{(\gamma+1)}\delta b_{m,n}(z)
\\[2em]
\begin{aligned}
&\sum_{\bar{m}=0}^{\widetilde{M}}\sum_{\bar{n}=0}^{\widetilde{N}}\frac{\alpha_{ia}\gamma C_{m,0,\bar{m},n,0,\bar{n}}}{4\pi^2(\gamma+1)}\delta a_{\bar{m},\bar{n}}(z)\\
&\quad-(\bar{b}_{m,n}-2\gamma\bar{a}_{m,n})\delta a_{0,0}(z)+\frac{\gamma}{\gamma+1}\delta a_{m,n}(z)
\end{aligned}
&
\begin{aligned}
&\sum_{\bar{m}=0}^{\widetilde{M}}\sum_{\bar{n}=0}^{\widetilde{N}}\frac{\alpha_{aa}\gamma(C_{m,\bar{m},0,n,\bar{n},0}+C_{m,0,\bar{m},n,0,\bar{n}})+\alpha_{ia}C_{m,\bar{m},0,n,\bar{n},0}}{4\pi^2(\gamma+1)}\delta b_{\bar{m},\bar{n}}(z)\\
&\quad-\frac{1+(m+2n)\tilde{\nu}_1}{\gamma+1}\delta b_{m,n}(z)
\end{aligned}
\end{pmatrix}
\tag{5.36}$$

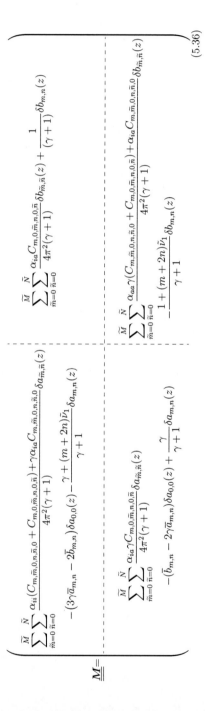

$$
\underline{\underline{L}} =
\left(
\begin{array}{c|c}
\begin{array}{l}
\beta(\gamma+1)\sqrt{m}\bar{a}_{m-1,n}\delta a_{0,0}(z) \\
+\sqrt{m}\delta a_{m-1,n}(z) \\
+\sqrt{m+1}\delta a_{m+1,n}(z)
\end{array}
& 0 \\
\hline
0 &
\begin{array}{l}
\sqrt{m}\delta b_{m-1,n}(z) \\
+\sqrt{m+1}\delta b_{m+1,n}(z)
\end{array}
\end{array}
\right)
\tag{5.37}
$$

Each of the 4 submatrices of $\underline{\underline{L}}$ and $\underline{\underline{M}}$ (indicated by dashed lines) has the dimension $(M+1)(N+1) \times (M+1)(N+1)$. The unknown functions $\delta a_{m,n}(z)$ and $\delta b_{m,n}(z)$ are not part of the actual matrices, but of the vector $\boldsymbol{f}$. They are only included in the symbolic forms above to demonstrate which coefficient in the matrices belongs to which function $\delta a_{m,n}(z)$ or $\delta b_{m,n}(z)$.

The problem given by equation (5.35) can be solved by an ansatz of the form $e^{\lambda z}$ for the functions $\delta a_{m,n}(z)$ and $\delta b_{m,n}(z)$, with λ representing the generalized eigenvalues of $\underline{\underline{M}}$ with respect to $\underline{\underline{L}}$ (see for example [39] - [41]). With a total of $S = 2(M+1)(N+1)$ eigenvalues λ_s and the corresponding eigenvectors $\boldsymbol{f}_s$, each combination satisfies the eigenvalue equation

$$
\underline{\underline{L}} \cdot \frac{\mathrm{d}\boldsymbol{f}_s}{\mathrm{d}z} = \underline{\underline{M}} \cdot \boldsymbol{f}_s = \lambda_s \underline{\underline{L}} \cdot \boldsymbol{f}_s
\tag{5.38}
$$

and the solution vector $\boldsymbol{f}$ can be written as a linear combination of the eigensystem:

$$
\boldsymbol{f} = \sum_{s=1}^{S} C_s \boldsymbol{f}_s e^{\lambda_s z}
\tag{5.39}
$$

Although $\underline{\underline{L}}$ and $\underline{\underline{M}}$ are not Hermitian, the eigenvalues are found to be real and come in pairs of one positive and one negative eigenvalue of the same absolute value. This is in agreement with the system's symmetry in z with respect to $z = 0$. There also exist two eigenvalues with value zero, representing the system's ambiguity with respect to a change in density or pressure, as mentioned above. It is helpful to make the system unambiguous by replacing equation (5.34) for $\{m = 0, n = 0\}$ and $\{m = 1, n = 0\}$ with the flux and pressure balances, equations (4.37) and (4.40). Since the replacement equations are algebraic, the result is a reduced differential equation system:

5. Solution for the Kinetic Model

$$\underline{\underline{L}}_{\text{red}} \cdot \frac{\mathrm{d}\boldsymbol{f}}{\mathrm{d}z} = \underline{\underline{M}}_{\text{red}} \cdot \boldsymbol{f} \tag{5.40}$$

The reduced matrices $\underline{\underline{L}}_{\text{red}}$ and $\underline{\underline{M}}_{\text{red}}$ are not square matrices anymore, which is however a prerequisite for the eigenvalue formalism. Thus, the matrices need to be reduced further. More specifically, two columns need to be removed to counterbalance the two rows removed before. The obvious choices are the rows representing $\delta b_{0,0}$ and $\delta b_{1,0}$. Making use, again, of the flux and pressure balance equations, all non-zero entries in these two rows can be expressed as linear combinations of coefficients in different rows and then can be removed. Afterwards, the rows representing $\delta b_{0,0}$ and $\delta b_{1,0}$ contain only zeros and can be removed completely. The result is a square matrix equation again,

$$\underline{\underline{L}}_{\text{rd}} \cdot \frac{\mathrm{d}\boldsymbol{f}_{\text{rd}}}{\mathrm{d}z} = \underline{\underline{M}}_{\text{rd}} \cdot \boldsymbol{f}_{\text{rd}}, \tag{5.41}$$

with the reduced vector

$$\boldsymbol{f}_{\text{rd}} = (\delta a_{0,0}, \delta a_{1,0}, \dots, \delta a_{M-1,N}, \delta a_{M,N}, \delta b_{2,0}, \dots, \delta b_{M-1,N}, \delta b_{M,N})^{T}.$$

This reduced eigenvalue problem,

$$\underline{\underline{L}}_{\text{rd}} \cdot \frac{\mathrm{d}\boldsymbol{f}_{\text{rd},s}}{\mathrm{d}z} = \underline{\underline{M}}_{\text{rd}} \cdot \boldsymbol{f}_{\text{rd},s} = \lambda_s \underline{\underline{L}}_{\text{rd}} \cdot \boldsymbol{y}_{\text{rd},s}, \tag{5.42}$$

has the same eigenvalues as equation (5.38), minus the two zero values. Accordingly, the linear combination of solutions to the reduced equation can be written as

$$\boldsymbol{f}_{\text{rd}} = \sum_{s=1}^{S/2-1} C_s^+ \boldsymbol{f}_s^+ e^{+\lambda_s z} + \sum_{s=1}^{S/2-1} C_s^- \boldsymbol{f}_s^- e^{-\lambda_s z}. \tag{5.43}$$

The Saha boundary condition now requires that only exponentially decreasing solutions are allowed:

$$\sum_{s=1}^{S/2-1} C_s^+ \boldsymbol{f}_s^+ e^{+\lambda_s z} \overset{!}{=} 0 \tag{5.44}$$

Since all eigenvectors belong to distinct eigenvalues, they are linearly independent and the condition above can only be satisfied when

$$C_s^+ e^{+\lambda_s z} = 0 \quad \text{for each } s = 1 \ldots S/2 - 1, \tag{5.45}$$

which basically means that all C_s^+ have to be zero. Accordingly, the condition by itself is of little help. However, the coefficients C_s^+ and C_s^- are related, and by investigating the left side eigenvalue problem related to equation (5.41), one can extract more information about the coefficients C_s^-. To this aim, the related left side eigenvalue problem is defined by

$$\tilde{\boldsymbol{f}}_{\text{rd},s} \cdot \underline{\underline{\boldsymbol{M}}}_{\text{rd}} = \lambda_s \tilde{\boldsymbol{f}}_{\text{rd},s} \cdot \underline{\underline{\boldsymbol{L}}}_{\text{rd}}. \tag{5.46}$$

Equations (5.42) and (5.46) have the same eigenvalues, and $\tilde{\boldsymbol{f}}_{\text{rd},s}$ and $\boldsymbol{f}_{\text{rd},s}$ are orthogonal (not necessarily orthonormal) with respect to $\underline{\underline{\boldsymbol{L}}}_{\text{rd}}$:

$$\tilde{\boldsymbol{f}}_{\text{rd},\sigma} \cdot \underline{\underline{\boldsymbol{L}}}_{\text{rd}} \cdot \boldsymbol{f}_{\text{rd},s} = C_{\sigma,s} \delta_{\sigma,s} \quad \text{and} \quad C_{\sigma,s} = 1 \quad \text{when orthonormal} \tag{5.47}$$

Therefore, the effect of multiplying $\tilde{\boldsymbol{f}}_{\text{rd},\sigma}$ with $\boldsymbol{f}_{\text{rd}}$, which is a linear combination of all $\boldsymbol{f}_s^+$ and $\boldsymbol{f}_s^-$, is similar to the sifting property of the delta function:

$$\tilde{\boldsymbol{f}}_{\text{rd},\sigma} \cdot \underline{\underline{\boldsymbol{L}}}_{\text{rd}} \cdot \boldsymbol{y} = \tilde{\boldsymbol{f}}_{\text{rd},\sigma} \cdot \underline{\underline{\boldsymbol{L}}}_{\text{rd}} \cdot \left(\sum_{s=1}^{S/2-1} C_s^+ \boldsymbol{f}_s^+ e^{+\lambda_s z} + \sum_{s=1}^{S/2-1} C_s^- \boldsymbol{f}_s^- e^{-\lambda_s z} \right)$$
$$= C_{\sigma,s} C_\sigma^\pm e^{\pm\lambda_\sigma z} \tag{5.48}$$

Evaluation of this property for all $\tilde{\boldsymbol{f}}_{\text{rd},\sigma}$ that belong to positive eigenvalues yields

$$\tilde{\boldsymbol{f}}_{\text{rd},\sigma} \cdot \underline{\underline{\boldsymbol{L}}}_{\text{rd}} \cdot \boldsymbol{f}_{\text{rd}} = C_{\sigma,s} C_\sigma^+ e^{+\lambda_\sigma z} \quad \text{for all } \sigma = 1 \ldots S/2 - 1, \tag{5.49}$$

which needs to be equal to zero according to equation (5.45); the normalization constant $C_{\sigma,s}$ has no further influence. Equation (5.49) now provides $S/2 - 1 = (M+1)(N+1) - 1$ linear algebraic equations for the $2(M+1)(N+1) - 2$ coefficients in $\boldsymbol{f}_{\text{rd}}$:

$$\tilde{\boldsymbol{f}}_{\text{rd},\sigma} \cdot \underline{\underline{\boldsymbol{L}}} \cdot \boldsymbol{f}_{\text{rd}} \overset{!}{=} 0 \quad \text{for each } \sigma = 1 \ldots S/2 - 1 \tag{5.50}$$

To complete the boundary condition near the Saha plasma, the equilibrium values of $a_{m,n}$ and $b_{m,n}$ need to be re-added into f and $f_{\rm rd}$ according to equations (5.31) and (5.32).

As expected, the linearization around the Saha plasma yields only $S/2-1$ of the necessary S boundary equations. The information to determine the remaining unknown boundary conditions is contained in the sheath boundary condition on the left.

5.2.2. Space Charge Sheath at $z = 0$

The second boundary condition is given by the assumption of a space charge sheath at the left edge ($z = 0$) of the modeled region. A completely absorbing wall is assumed, which means that all ions enter the space charge sheath with a certain distribution and eventually reach the cathode where they are neutralized. These newly produced neutrals then make their way back through the space charge sheath and enter the region of interest with another distribution function. The only information carried over from the ion flux to the neutral flux is given by the mass conservation law. All other information (like energy and momentum of the ions) is lost during the conversion process. In the following sections, the details of this process will be discussed.

Neutrals

The neutrals coming out of the space charge sheath originate from two sources: i) a flux of neutrals to the cathode and ii) a flux of ions to the cathode. The details of the two processes are discussed further below. For now, the two-fold origin suggests expressing the neutral distribution function as a superposition of two other distribution functions:

$$f_a(v_{\|}, v_{\perp}^2)|_{z=0} = f_a^a(v_{\|}, v_{\perp}^2) + f_a^i(v_{\|}, v_{\perp}^2) \qquad (5.51)$$

Furthermore, the distribution functions can be divided into even and odd components, which will be helpful for further calculations:

$$f_a = f_a^e + f_a^o = f_a^{a,e} + f_a^{a,o} + f_a^{i,e} + f_a^{i,o} \qquad (5.52)$$
$$f_a^e = f_a^{a,e} + f_a^{i,e} \qquad (5.53)$$
$$f_a^o = f_a^{a,o} + f_a^{i,o} \qquad (5.54)$$

The explicit dependence of f on z, $v_{\|}$, and $v_{\perp}$ has been omitted for readability.

For the further investigation, f_a^a shall be considered first. The process generating f_a^a can be described as follows: Neutrals from the presheath enter the space charge sheath. They travel to the cathode completely undisturbed, since they do not carry a charge and since the space charge sheath is collision free. The neutral distribution function is zero for all $v_\parallel > 0$, since only neutrals with negative speeds at the sheath edge can enter. At the cathode, the neutrals coming from the plasma are reflected and travel back the way they came, again undisturbed, until they reenter the presheath. For the distribution function, the reflection is equivalent to mirroring the distribution function at $v_\parallel = 0$. Superposition of the original distribution function and its mirror eliminates all odd components, so that all odd components of f_a stem from f_a^i:

$$f_a^{a,o} = 0 \tag{5.55}$$
$$\Rightarrow \quad f_a^o = f_a^{i,o} \tag{5.56}$$

Before determining f_a^i further (and thereby f_a^o), it should be noted that assuming a simple reflection of neutrals at the cathode ensures particle conservation, but neglects the thermal coupling between plasma and electrode. However, the number of particles in this first group is rather small. Thus, the error caused by neglecting thermal coupling for reflected neutrals is small as well.

For the second group, represented by f_a^i, the number of particles is not small and thermal coupling needs to be taken into account. The neutrals in this second group originate from ions entering the space charge sheath from the presheath side. Ions are strongly accelerated within the sheath and reach the electrode with an almost beam-like distribution function. Simply reflecting them as neutrals from the electrode surface would cause significant errors. Instead, it is assumed that the ions enter the electrode and leave again as neutrals with temperature $T_{a,0}$ through thermal emission. This process ensures mass conservation and the value of $T_{a,0}$ allows a thermal coupling between the cathode (temperature T_c) and the adjacent plasma (temperature T_h). The closer $T_{a,0}$ gets to T_c, the stronger the coupling becomes.

Mathematically, the distribution function of thermally emitted neutrals can be assumed to be half Maxwellian: only neutrals with $v_\parallel > 0$ actually leave the cathode. With the temperature $T_{a,0} = T_h/\vartheta$ and the Heaviside step function,

$$\Theta(v_\parallel)_0 = \begin{cases} 0, & v_\parallel \le 0 \\ 1, & v_\parallel > 0 \end{cases} \tag{5.57}$$

the distribution function f_a^i with density n_{ac} can be written as:

5. Solution for the Kinetic Model

$$f_a^i(\boldsymbol{v}) = \Theta(v_\parallel) f_M(\boldsymbol{v}) = \Theta(v_\parallel) 2n_{ac} \left(\frac{\vartheta}{2\pi}\right)^{3/2} e^{-\vartheta\left(v_\parallel^2 + v_\perp^2\right)/2} \tag{5.58}$$

The unknown density n_{ac} can be calculated from the flux conservation: The first moment of f_a^i needs to be equal to the flux of ions entering the sheath. These ions are represented by the parts of f_i with negative $v_\parallel$, denoted by f_i^-:

$$\int v_\parallel f_a^i(\boldsymbol{v}) \mathrm{d}^3 v = -\int v_\parallel f_i^-(\boldsymbol{v}) \mathrm{d}^3 v \tag{5.59}$$

$$\Rightarrow \qquad \frac{2n_{ac}}{\sqrt{2\pi\vartheta}} = -2\pi \int_{-\infty}^0 \int_0^\infty v_\parallel f_i(v_\parallel, v_\perp^2) v_\perp \mathrm{d}v_\perp \mathrm{d}v_\parallel$$

$$\Rightarrow \qquad n_{ac} = -\frac{\sqrt{2\pi\vartheta}}{2} \int_{-\infty}^0 \int_0^{2\pi} \int_0^\infty v_\parallel f_i(v_\parallel, v_\perp^2) v_\perp \mathrm{d}v_\perp \mathrm{d}\varphi \mathrm{d}v_\parallel$$

$$= -\int_{-\infty}^0 \int_0^\infty v_\parallel \sum_{m=0}^M \sum_{n=0}^N \frac{\sqrt{\vartheta} a_{m,n}}{2} e^{-(v_\parallel^2 + v_\perp^2)/2} \mathrm{He}_m(v_\parallel) \mathrm{La}_n(v_\perp^2) v_\perp \mathrm{d}v_\perp \mathrm{d}v_\parallel$$

$$= -\int_{-\infty}^0 v_\parallel \sum_{m=0}^M \frac{\sqrt{\vartheta} a_{m,0}}{2} e^{-v_\parallel^2/2} \mathrm{He}_m(v_\parallel) \mathrm{d}v_\parallel \tag{5.60}$$

This integral can be solved with equations (A.31) and (A.32), so that one finds for n_{ac}:

$$n_{ac} = \frac{\sqrt{\vartheta}}{2} \left[a_{0,0} e^{-v_\parallel^2/2} + a_{1,0} \left(v_\parallel e^{-v_\parallel^2/2} - \sqrt{\frac{\pi}{2}} \mathrm{erf}\left(\frac{v_\parallel}{\sqrt{2}}\right) \right) \right.$$

$$\left. - \sum_{m=2}^M \frac{(-1)^m a_{m,0}}{\sqrt{m!}} e^{-v^2/2} \left(\sum_{i=0}^{m-1} h_{m-1,i}^- v^{i+1} - \sum_{j=0}^{m-2} h_{m-2,j}^- v^j \right) \right]_{v_\parallel=-\infty}^{v_\parallel=0} \tag{5.61}$$

The combination of equations (5.56), (5.58), and (5.60) yields a boundary condition for the odd part of the neutral distribution function. All results can be transferred directly to the coefficients by application of the expansion, the even/odd characteristics of the Hermite polynomials, and the standard rules for combining even and odd functions (all considerations are made at $z = 0$):

$$f_a(v_\parallel, v_\perp^2) = \sum_{m=0}^M \sum_{n=0}^N b_{m,n} \mathrm{He}_m(v_\parallel) \mathrm{La}_n(v_\perp^2) e^{-(v_\parallel^2 + v_\perp^2)/2} \tag{5.62}$$

$$
\Rightarrow \begin{cases}
f_a^{(a/i)}(v_\parallel, v_\perp^2) = \sum_{m=0}^{M} \sum_{n=0}^{N} b_{m,n}^{(a/i)} \mathrm{He}_m(v_\parallel) \mathrm{La}_n(v_\perp^2) e^{-(v_\parallel^2 + v_\perp^2)/2} \\[3ex]
f_a^{(a/i),e}(v_\parallel, v_\perp^2) = \sum_{m=0}^{(M-1)/2} \sum_{n=0}^{N} b_{2m,n}^{(a/i)} \mathrm{He}_{2m}(v_\parallel) \mathrm{La}_n(v_\perp^2) e^{-(v_\parallel^2 + v_\perp^2)/2} \\[3ex]
f_a^{(a/i),o}(v_\parallel, v_\perp^2) = \sum_{m=0}^{(M-1)/2} \sum_{n=0}^{N} b_{2m+1,n}^{(a/i)} \mathrm{He}_{2m+1}(v_\parallel) \mathrm{La}_n(v_\perp^2) e^{-(v_\parallel^2 + v_\perp^2)/2}
\end{cases}
$$

From this equation and equation (5.56), one finds

$$
f_a^o = f_a^{i,o}
$$

$$
\Rightarrow \sum_{m=0}^{(M-1)/2} \sum_{n=0}^{N} b_{2m+1,n} \mathrm{He}_{2m+1}(v_\parallel) \mathrm{La}_n(v_\perp^2) e^{-(v_\parallel^2 + v_\perp^2)/2}
$$
$$
= \sum_{m=0}^{(M-1)/2} \sum_{n=0}^{N} b_{2m+1,n}^i \mathrm{He}_{2m+1}(v_\parallel) \mathrm{La}_n(v_\perp^2) e^{-(v_\parallel^2 + v_\perp^2)/2}. \tag{5.63}
$$

Finally, due to the orthogonality of the Hermite polynomials, the equality above needs to be met for each coefficient of the sum by itself:

$$
b_{m,n} = b_{m,n}^i \quad \text{if } m \text{ is odd} \tag{5.64}
$$

The coefficients $b_{m,n}^i$ can be calculated from equations (4.66) and (5.58):

$$
b_{m,n}^i = \int_0^\infty \int_0^\infty \int_0^{2\pi} 2 n_{ac} \left(\frac{\vartheta}{2\pi} \right)^{3/2} e^{-\vartheta(v_\parallel^2 + v_\perp^2)/2} \mathrm{He}_m(v_\parallel) \mathrm{La}_n(v_\perp^2) v_\perp \, d\varphi \, dv_\perp \, dv_\parallel
$$
$$
= 2 n_{ac} \int_0^\infty \int_0^\infty \frac{\vartheta^{3/2}}{\sqrt{2\pi}} e^{-\vartheta(v_\parallel^2 + v_\perp^2)/2} \mathrm{He}_m(v_\parallel) \mathrm{La}_n(v_\perp^2) v_\perp \, dv_\perp \, dv_\parallel \tag{5.65}
$$

Although this integral cannot be evaluated by an orthogonality rule, it can be calculated analytically for any combination of m and n. The results are polynomial fractions in $\sqrt{\vartheta}$ and n_{ac}, exemplary results for up to $m = 3$ and $n = 2$ are shown in table 5.1; coefficients with even m are included for completeness.

	n=0	1	2
m=0	n_{ac}	$\dfrac{n_{ac}(1-\vartheta)}{\vartheta}$	$\dfrac{n_{ac}(1-\vartheta)^2}{\vartheta^2}$
1	$\dfrac{2n_{ac}}{\sqrt{2\pi\vartheta}}$	$\dfrac{2n_{ac}(1-\vartheta)}{\sqrt{2\pi}\vartheta^{3/2}}$	$\dfrac{2n_{ac}(1-\vartheta)^2}{\sqrt{2\pi}\vartheta^{5/2}}$
2	$\dfrac{n_{ac}(1-\vartheta)}{\sqrt{2}\vartheta}$	$\dfrac{n_{ac}(1-\vartheta)^2}{\sqrt{2}\vartheta^2}$	$\dfrac{n_{ac}(1-\vartheta)^3}{\sqrt{2}\vartheta^3}$
3	$\dfrac{n_{ac}(2-3\vartheta)}{\sqrt{3\pi}\vartheta^{3/2}}$	$\dfrac{n_{ac}(1-\vartheta)(2-3\vartheta)}{\sqrt{3\pi}\vartheta^{5/2}}$	$\dfrac{n_{ac}(1-\vartheta)^2(2-3\vartheta)}{\sqrt{3\pi}\vartheta^{7/2}}$

Table 5.1.: Results of $b^i_{m,n}$ according to definition (5.65) for up to $m=3$ and $n=2$.

With n_{ac} from equation (5.60) equation (5.65) yields an expression for all $b^i_{m,n}$ in dependence on the coefficients $a_{m,n}$. Since for odd m the coefficients $b^i_{m,n}$ are equivalent to the coefficients $b_{m,n}$, equations (5.60) and (5.65) yield a boundary condition for half of the coefficients $b_{m,n}$ (i.e., the ones with odd m).

Ions

All ions entering the sheath have a negative velocity. They travel through the space charge sheath without collisions and are constantly accelerated towards the electrode by the strong electric field in the space charge sheath. At the electrode, the ions are absorbed and neutralized. Accordingly, there exist no ions with positive velocity throughout the space charge sheath, including its border at $z = 0$. In terms of the ion distribution function, this condition is simply expressed by:

$$f(0, v_\parallel, v_\perp^2) \equiv 0 \quad \text{for} \quad v_\parallel \geq 0 \tag{5.66}$$

To use this condition with the orthogonal expansion, the definition of the coefficients $a_{m,n}$ is recalled (equation (4.65)) and considered at $z = 0$:

$$a_{m,n}(0) = \int_{-\infty}^{\infty} \int_0^{\infty} \int_0^{2\pi} f_i(0, v_\parallel, v_\perp^2) \text{He}_m(v_\parallel) \text{La}_n(v_\perp^2) v_\perp \, d\varphi dv_\perp dv_\parallel \tag{5.67}$$

The definiton of the expansion itself (equation (4.63)) is resubstituted into this definition.

The orthogonality criterion is applied only with respect to $v_\perp^2$ and the boundary condition (5.66) is applied for $v_\parallel$ instead:

$$
\begin{aligned}
a_{m,n}(0) &= \int_{-\infty}^{\infty} \int_{0}^{2\pi} \int_{0}^{\infty} f_i(0, v_\parallel, v_\perp^2) \mathrm{He}_m(v_\parallel) \mathrm{La}_n(v_\perp^2) v_\perp \, \mathrm{d}v_\perp \, \mathrm{d}\varphi \, \mathrm{d}v_\parallel \\
&= \int_{-\infty}^{0} \sum_{\tilde{m}=0}^{M} \frac{a_{\tilde{m},n}}{\sqrt{2\pi}} \mathrm{He}_{\tilde{m}} e^{-v_\parallel^2/2} \mathrm{He}_m(v_\parallel) \, \mathrm{d}v_\parallel \\
&= \int_{-\infty}^{0} \sum_{\tilde{m}=0}^{M} \frac{a_{\tilde{m},n}}{\sqrt{2\pi}} \sum_{k_1=0}^{[\tilde{m}/2]} \frac{(-1)_1^k \sqrt{\tilde{m}!}}{k_1! 2_1^k (\tilde{m}-2k_1)!} v^{\tilde{m}-2k_1} e^{-v_\parallel^2/2} \sum_{k=0}^{[m/2]} \frac{(-1)^k \sqrt{m!}}{k! 2^k (m-2k)!} v^{m-2k} \, \mathrm{d}v_\parallel \\
&= \sum_{\tilde{m}=0}^{M} \sum_{k_1=0}^{[\tilde{m}/2]} \sum_{k=0}^{[m/2]} \frac{a_{\tilde{m},n} (-1)^{k+k_1} \sqrt{m! \tilde{m}!}}{\sqrt{2\pi} k! k_1! 2^{k+k_1} (m-2k)! (\tilde{m}-2k_1)!} \int_{-\infty}^{0} v^{m+\tilde{m}-2k-2k_1} e^{-v_\parallel^2/2} \, \mathrm{d}v_\parallel \\
&= \sum_{\tilde{m}=0}^{M} \sum_{k_1=0}^{[\tilde{m}/2]} \sum_{k=0}^{[m/2]} \frac{a_{\tilde{m},n} (-1)^{m+\tilde{m}+k+k_1} 2^{(m+\tilde{m}-1)/2} \sqrt{m! \tilde{m}!}}{\sqrt{2\pi} k! k_1! 2^{2k+2k_1} (m-2k)! (\tilde{m}-2k_1)!} \Gamma\left(\frac{m+\tilde{m}-2k-2k_1+1}{2}\right)
\end{aligned}
$$

$$(5.68)$$

Equations (A.21) and (A.58) were used in the course of the calculation above. The result can be understood as a relation among all coefficients $a_{m,n}$. For $M \to \infty$ this relation ensures that the distribution function f_i is zero for all $v_\parallel \geq 0$. For all finite values of M there exists a certain error which decreases with increasing M. Oscillations like Gibb's phenomenon do not occur, since f_i is steady for all $v_\parallel$.

Together with equations (5.29) and (5.30) and the first boundary condition (equation (5.49)), this poses a non-linear boundary value problem for $2(M+1)(N+1)$ variables.

In chapter 3, a similar boundary problem for two variables was reduced to an initial value problem with one degree of freedom. The starting point was taken as a point close to the Saha plasma, and the corresponding equations were solved progressively towards the cathode. Since there were only two variables (v_i and v_a, from which n_i and n_a were calculated algebraically), every solution satisfied the Bohm criterion at some point, and the remaining degree of freedom simply determined the distance from the starting point to the Bohm point. Such a procedure is not appropriate for this new boundary problem: the number of variables is higher, and not every solution starting near the Saha plasma necessarily passes through the Bohm point. In fact, the vast majority of solutions does not pass through it, thus the boundary value problem here cannot be reduced to an initial value problem. Instead, the boundary problem will be solved by a relaxation method. This method requires a discretization of the differential equations (5.29) and (5.30), as well as the boundary conditions (5.65) and (5.66).

5.3. Discretization

5.3.1. The Grid Equations

Since the equations and boundary conditions derived in the sections above are highly non-linear, they can only be solved numerically. The procedures used will be described in this section. They are based on the finite volume method, which is well-suited for conservation equations like the Boltzmann equation.

The problem that needs to be solved here can be described as a boundary problem for $S = 2(M+1)(N+1)$ one-dimensional functions $a_{m,n}(z)$ and $b_{m,n}(z)$ on a given interval $[0, L]$, with $L > 0$. The problem is defined by $S-2$ differential and 2 algebraic equations, plus a set of boundary conditions at $z = 0$ and $z = L$. A uniform grid with K points is defined over the interval, as demonstrated in figure 5.1.

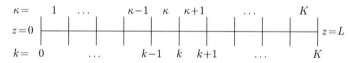

Figure 5.1.: Diagram of the discretization. The $K + 1$ grid points (index k) define K subintervals/cells (index κ) on the interval $0 \leq z \leq L$.

The grid constant is $h = L/K$. The grid point positions are defined by $z_k = kh$. They define subintervals/cells $\overline{z}_\kappa$, with κ ranging from 1 to K. On a given subinterval, the functions $a_{m,n}(z)$ and $b_{m,n}(z)$ are represented by constants equal to the average function value on the corresponding interval:

$$a_{m,n}(z) \;:=\; \overline{a}_{m,n,\kappa} = \frac{1}{z_k - z_{k-1}} \int_{z_{k-1}}^{z_k} a_{m,n}(z)\mathrm{d}z, \quad \text{for} \quad z_{k-1} < z < z_k \quad (5.69)$$

$$b_{m,n}(z) \;:=\; \overline{b}_{m,n,\kappa} = \frac{1}{z_k - z_{k-1}} \int_{z_{k-1}}^{z_k} b_{m,n}(z)\mathrm{d}z, \quad \text{for} \quad z_{k-1} < z < z_k \quad (5.70)$$

At the inner grid points, the function values are interpolated from adjacent cells:

$$a_{m,n}(z)|_{z=z_k} \;:=\; a_{m,n,k} = \frac{\overline{a}_{m,n,\kappa+1} + \overline{a}_{m,n,\kappa}}{2} \quad (5.71)$$

$$b_{m,n}(z)|_{z=z_k} \;:=\; b_{m,n,k} = \frac{\overline{b}_{m,n,\kappa+1} + \overline{b}_{m,n,\kappa}}{2} \quad (5.72)$$

Values for the outer grid points ($k = 0$, $k = K$) are obtained from boundary conditions and extrapolation, which will be discussed further below. Application of the averaging integral to the derivatives leads to the difference of the function values at neighboring grid points:

$$\frac{1}{z_k - z_{k-1}} \int_{z_{k-1}}^{z_k} a'_{m,n}(z)\mathrm{d}z = \frac{a_{m,n,k} - a_{m,n,k-1}}{z_k - z_{k-1}} = \frac{\overline{a}_{m,n,\kappa+1} + \overline{a}_{m,n,\kappa-1}}{2(z_k - z_{k-1})} \qquad (5.73)$$

$$\frac{1}{z_k - z_{k-1}} \int_{z_{k-1}}^{z_k} b'_{m,n}(z)\mathrm{d}z = \frac{b_{m,n,k} - b_{m,n,k-1}}{z_k - z_{k-1}} = \frac{\overline{a}_{m,n,\kappa+1} + \overline{a}_{m,n,\kappa-1}}{2(z_k - z_{k-1})} \qquad (5.74)$$

Carrying out the replacements above transforms the S original differential equations into a set of KS algebraic equations. For the inner cells ($\kappa = 2 \ldots K - 1$), they read as follows:

$$
\begin{aligned}
&\sqrt{m}\frac{(\overline{a}_{m-1,n,\kappa+1} - \overline{a}_{m-1,n,\kappa-1})}{2(z_k - z_{k-1})} + \sqrt{m+1}\frac{(\overline{a}_{m+1,n,\kappa+1} - \overline{a}_{m+1,n,\kappa-1})}{2(z_k - z_{k-1})} \\
&+ \frac{\sqrt{m}\beta}{\overline{a}_{0,0,\kappa}}\frac{(\overline{a}_{0,0,\kappa+1} - \overline{a}_{0,0,\kappa-1})}{2(z_k - z_{k-1})}\overline{a}_{m-1,n,\kappa} \\
&= \overline{a}_{0,0,\kappa}\overline{b}_{m,n,\kappa} - \gamma(1+\gamma)\overline{a}_{0,0,\kappa}^2\overline{a}_{m,n,\kappa} - \tilde{\nu}_1(2n+m)\overline{a}_{m,n,\kappa} \\
&+ \sum_{\tilde{m}=0}^{\overline{M}}\sum_{\tilde{n}=0}^{\overline{N}}\sum_{\tilde{\tilde{m}}=0}^{\tilde{\tilde{M}}}\sum_{\tilde{\tilde{n}}=0}^{\tilde{\tilde{N}}} \frac{\alpha_{ii}\overline{a}_{\tilde{m},\tilde{n},\kappa}\overline{a}_{\tilde{\tilde{m}},\tilde{\tilde{n}},\kappa} + \alpha_{ia}\overline{a}_{\tilde{m},\tilde{n},\kappa}\overline{b}_{\tilde{\tilde{m}},\tilde{\tilde{n}},\kappa}}{4\pi^2}C_{m,n,\tilde{m},\tilde{\tilde{m}},\tilde{n},\tilde{\tilde{n}}}
\end{aligned}
\qquad (5.75)
$$

$$
\begin{aligned}
&\sqrt{m}\frac{(\overline{b}_{m-1,n,\kappa+1} - \overline{b}_{m-1,n,\kappa-1})}{2(z_k - z_{k-1})} + \sqrt{m+1}\frac{(\overline{b}_{m+1,n,\kappa+1} - \overline{b}_{m+1,n,\kappa-1})}{2(z_k - z_{k-1})} \\
&= -\overline{a}_{0,0,\kappa}\overline{b}_{m,n,\kappa} + \gamma(1+\gamma)\overline{a}_{0,0,\kappa}^2\overline{a}_{m,n,\kappa} - \tilde{\nu}_1(2n+m)\overline{b}_{m,n,\kappa} \\
&+ \sum_{\tilde{m}=0}^{\overline{M}}\sum_{\tilde{n}=0}^{\overline{N}}\sum_{\tilde{\tilde{m}}=0}^{\tilde{\tilde{M}}}\sum_{\tilde{\tilde{n}}=0}^{\tilde{\tilde{N}}} \frac{\alpha_{aa}\overline{b}_{\tilde{m},\tilde{n},\kappa}\overline{b}_{\tilde{\tilde{m}},\tilde{\tilde{n}},\kappa} + \alpha_{ai}\overline{b}_{\tilde{m},\tilde{n},\kappa}\overline{a}_{\tilde{\tilde{m}},\tilde{\tilde{n}},\kappa}}{4\pi^2}C_{m,n,\tilde{m},\tilde{\tilde{m}},\tilde{n},\tilde{\tilde{n}}}
\end{aligned}
\qquad (5.76)
$$

The four equations for the outer cells ($k = 0$, $k = K$) have the same structure; however, they still contain the grid point values $a_{m,n,0}$, $b_{m,n,0}$, $a_{m,n,K}$, and $b_{m,n,K}$. These grid point values cannot be determined from interpolation, but need to be treated as additional unknowns. Accordingly, one has $(K+2)S$ unknown variables but only KS grid equations. To find a unique solution (if it exists), $S(K + 2)$ equations are needed. The missing $2S$ equations need to be determined from the boundary conditions instead. This will be discussed in the next two sections. Before this discussion, it is worth noting that the approach above is almost identical to a simple discretization with a centralized difference quotient. However, the more general formulation above is more convenient for possible improvements of the discretization (e.g., a non-uniform grid or higher-order function approximations within the cells, see [42] for more details).

5.3.2. Boundary Conditions: Saha Plasma at $z \to \infty$

This and the next section describe how the boundary conditions from sections (5.2.1) and (5.2.2) are formulated in discretized form. Since the original system of differential equations consists of $S = 2(M + 1)(N + 1)$ first-order differential equations, it is fully determined by S boundary conditions. In contrast, it was found above that $2S$ equations are missing. The discrepancy between these two numbers is due to the discretization scheme and does not represent missing physical information. It is therefore sufficient to find the S equations from the boundary conditions and to formally obtain S equations by extrapolation. These extrapolated equations do not hold additional physical information.

Vector equation (5.50) in chapter 5.2.1 represents a set of $S/2 - 1$ boundary conditions for the (equilibrium-free) coefficients $\delta a_{m,n}(z)$ and $\delta b_{m,n}(z)$. Since the Saha plasma is characterized by a Maxwellian distribution function, the equilibrium values are given by

$$a_{m,n}^{\mathrm{eq}} = \frac{1}{1 + \gamma}\delta_{m,0}\delta_{n,0} \quad \text{and} \quad b_{m,n}^{\mathrm{eq}} = \frac{\gamma}{1 + \gamma}\delta_{m,0}\delta_{n,0}, \tag{5.77}$$

with the Kronecker delta $\delta_{n,\tilde{n}}$. Re-adding these equilibrium values according to

$$a_{m,n}(z) = a_{m,n}^{\mathrm{eq}} + \delta a_{m,n}(z) \tag{5.78}$$
$$b_{m,n}(z) = b_{m,n}^{\mathrm{eq}} + \delta b_{m,n}(z) \tag{5.79}$$

and discretizing the equations according to equations (5.69) and (5.70) results in $S/2 - 1$ boundary conditions for the discretized coefficients $a_{m,n,\kappa}$ and $b_{m,n,\kappa}$. Two more equations can be obtained from discretizing the conservation equations (5.1) and (5.2), which are also valid in the Saha plasma. Thus, $S/2 + 1$ physical equations are obtained from the right side boundary conditions. The missing $S/2 - 1$ equations are obtained from extrapolation. This extrapolation can be linear, since the right boundary should be chosen near the Saha plasma, where the exponential behavior is already strongly saturated.

5.3.3. Boundary Conditions: Space Charge Sheath Edge at $z = 0$

The boundary equations on the left side can be found from equations (5.65) and (5.68) by replacing the coefficients $a_{m,n}(0)$ and $b_{m,n}(0)$ with their discrete counterparts $a_{m,n,0}$ and $b_{m,n,0}$. The equation stemming from the reflected neutrals then reads:

$$b_{m,n,0} = \int_0^\infty \int_0^\infty n_{ac} \frac{\vartheta^{3/2}}{\sqrt{2\pi}} e^{-\vartheta\left(v_\parallel^2 + v_\perp^2\right)/2} \mathrm{He}_m(v_\parallel) \mathrm{La}_n(v_\perp^2) v_\perp \mathrm{d}v_\perp \mathrm{d}v_\parallel, \quad \text{if } m \text{ is odd,} \quad (5.80)$$

with

$$n_{ac} = \sqrt{\vartheta} \left[a_{0,0,0} e^{-v_\parallel^2/2} + a_{1,0,0} \left(v_\parallel e^{-v_\parallel^2/2} - \sqrt{\frac{\pi}{2}} \mathrm{erf}\left(\frac{v_\parallel}{\sqrt{2}}\right) \right) \right.$$
$$\left. - \sum_{m=2}^M \frac{(-1)^m a_{m,0,0}}{\sqrt{m!}} e^{-v^2/2} \left(\sum_{i=0}^{m-1} h_{m-1,i}^- v^{i+1} - \sum_{j=0}^{m-2} h_{m-2,j}^- v^j \right) \right]_{v_\parallel=-\infty}^{v_\parallel=0} \quad (5.81)$$

Equation 5.80 is given only for instances of $b_{m,n}$ with odd m, thus it represents $S/4$ physical boundary conditions. Accordingly, $S/4 - 1$ physical boundary equations need to be extracted from equation (5.68):

$$a_{m,n,0} = \sum_{\tilde{m}=0}^M \sum_{k_1=0}^{[\tilde{m}/2]} \sum_{k=0}^{[m/2]} \frac{a_{\tilde{m},n,0}(-1)^{m+\tilde{m}+k+k_1} 2^{(m+\tilde{m}-1)/2} \sqrt{m!\tilde{m}!}}{\sqrt{2\pi} k! k_1! 2^{2k+2k_1} (m-2k)! (\tilde{m}-2k_1)!} \Gamma\left(\frac{m+\tilde{m}-2k-2k_1+1}{2}\right)$$
$$(5.82)$$

The decision which combinations of m and n to choose in equation (5.82) is arbitrary, as long as the total number of combinations is equal to $S/4 - 1$. Picking more (fewer) combinations would lead to an overdetermined (underdetermined) equation system. Just as in the previous section, the technically necessary equations remaining are, again, obtained by extrapolation. When extrapolating coefficients $b_{m,n}$, a linear extrapolation should be used, whereas an extrapolation for $a_{m,n}$ should assume a square-root like behavior. This can be found from investigating the coefficient behavior in proximity of the Bohm point, see section A.2.

5.4. Results

This section contains solutions obtained from the kinetic model introduced above. The purpose is to show what typical solutions look like and that the kinetic model is an actual improvement compared to the fluid model. The solutions were calculated by numerically solving the equations deduced above with Wolfram Mathematica. The command used (*FindRoot[]*) uses a damped Newton's method. Due to the discretization scheme and the large number of variables, this method leads to some ripples or oscillations in the solutions. These anomalies can be suppressed, for example, by using a smaller damping factor (i.e. stronger damping), which leads to slower convergence. The following plots (cf. figure 5.2) were calculated for $K = 100$, $M = 7$, $N = 1$, and with a damping factor that yields a reasonable compromise between computation speed and suppression of oscillations. With $K = 100$, the resulting number of variables to solve for is $2(K + 1)(M + 1)(N + 1) = 3,232$.

It can be observed that the absolute maximum value of the coefficients decreases quickly with increasing m and even quicker with increasing n. This means that the error caused by using finite upper limits M and N also increases quickly. In the particular case shown in figure 5.2, an upper limit of $M = 3$ would have been sufficient. The oscillations described above can also be seen in the plots. By applying equations (5.71) and (5.72), it is possible to calculate the function values at the grid points. When plotting these instead of the average cell values, hardly any oscillations are observed, as can be seen in figure 5.3. For the plots in examples I - IV, this kind of plot is used.

Combining all coefficients according to equations 4.63 and 4.64 yields the complete heavy particle distribution functions $f_i(z)$ and $f_a(z)$, respectively. These distribution functions contain much more information about the particle behavior than just the density and velocity information obtained from fluid models. This knowledge also allows one to plot the distribution function in velocity space. Figures 5.4 and 5.5 show the typical transit behavior of the ion distribution function from the sheath edge towards the Saha plasma.

In order to compare the kinetic results with the results obtained from the fluid models in chapters 2 and 3, the relation between the parameters used in each model need to be known. The parameters in the fluid models are α, β, and γ. (With γ always equal to zero for the original model.) The two parameters β and γ are identical to β and γ in the kinetic model. α has been split into α_{ii}, α_{ia}, α_{ai}, and α_{aa} in the kinetic model. It is reasonable to always set $\alpha_{ia} = \alpha_{ai}$ since the cross section for an ion-atom collision is normally the same as the cross section for an atom-ion collision. The ion-ion and atom-atom cross sections are usually different but are assumed to be equal in this section. This reduces the number of four α-parameters in the kinetic model to a single α_{kin}:

$$\alpha_{\text{kin}} = \alpha_{ii} = \alpha_{aa} = \alpha_{ia} = \alpha_{ai} \tag{5.83}$$

The relation between this α_{kin} and α in the fluid dynamic models can be obtained by taking into account the original definition of α and its dependence on the cross section (see [28], p. 3109):

$$\alpha_{\text{kin}} = \frac{3\alpha^2}{32\sqrt{\pi}} \tag{5.84}$$

The parameter ν_1 is not available in the fluid model. It is a replacement for the radial loss of energy, which cannot be realized in the model due to the one-dimensional approximation. The way that ν_1 was introduced, it is equivalent to a thermal coupling of the heavy particles to a pool of particles at temperature T_h. Different values for ν_1 influences the length scale of the transition to the Saha plasma. Smaller values of ν_1 (equivalent to a weak thermal coupling) enlarge the length scale, whereas larger values of ν_1 (equivalent to a strong thermal coupling) decrease the length scale. This influence is, however, rather weak and can only be noticed for very small ($< 10^{-3}$) or very large (> 10) values of ν_1. For the following examples, ν_1 is held constant at 1.

Finally, the parameter ϑ is also held constant at 1 for now to focus on the influence that α and γ have on the results. Physically, this means that the cathode is thermally not coupled to the plasma.

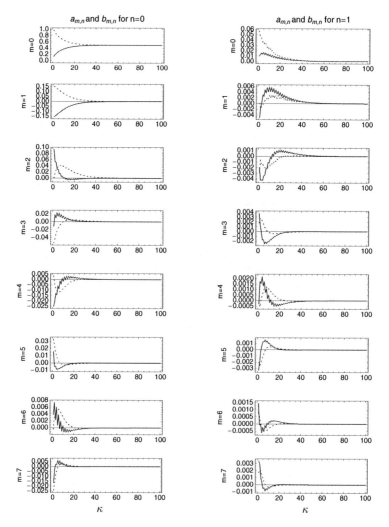

Figure 5.2.: Exemplary solution plots for all coefficients $a_{m,n}$ (solid) and $b_{m,n}$ (dashed) for $m = 0 \ldots 7$ and $n = 0 \ldots 1$ over cell index κ. The left column keeps $n = 0$ with m ascending from top to bottom; the right column keeps $n = 1$ with m ascending from top to bottom.

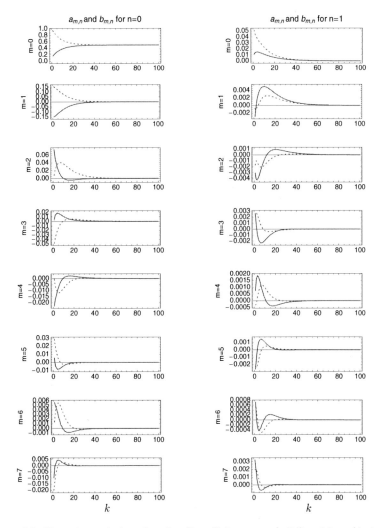

Figure 5.3.: Exemplary solution plots for all coefficients $a_{m,n}$ (solid) and $b_{m,n}$ (dashed) for $m = 0 \ldots 7$ and $n = 0, 1$ over grid point index k. The left column keeps $n = 0$ with m ascending from top to bottom; the right column keeps $n = 1$ with m ascending from top to bottom.

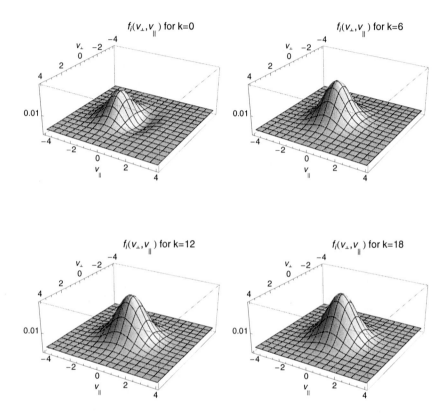

Figure 5.4.: Transition of the ion distribution function in the $v_\parallel - v_\perp$−space from the Bohm edge (left, $k = 0$) to the Saha plasma (right, $k \gtrsim 18$). The small ripples around $v_\parallel = 0$ for $k = 0$ stem from the finite limit M of the expansion and numerical errors in the coefficient calculation.

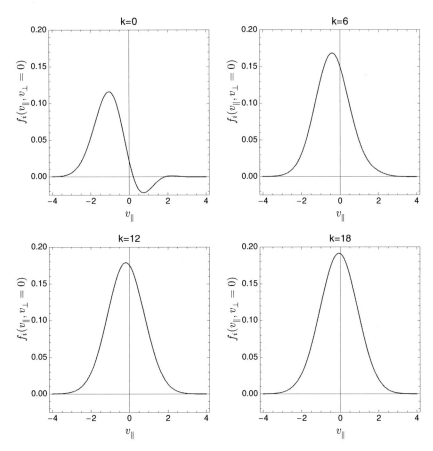

Figure 5.5.: Transition of the marginal ion distribution function in the $v_\parallel$−space from the Bohm edge (left, $k = 0$) to the Saha plasma (right, $k \gtrsim 18$). The small ripples around $v_\parallel = 0$ for $k = 0$ stem from the finite limit M of the expansion and numerical errors in the coefficient calculation.

5. Solution for the Kinetic Model

5.4.1. Example I: $\{\alpha = 2, \beta = 1, \gamma = 1\}$

This example describes the least critical situation, with $\alpha = 2 > 1$ and $\gamma = 1 > 0$. The results (figures 5.6 and 5.7) are in good agreement with the results from the expanded fluid model.

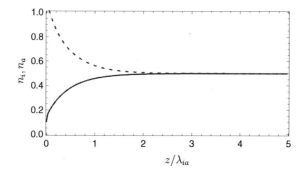

Figure 5.6.: Densities for the unique solution. The solid line shows the ion density, the dashed line shows the neutral density ($\{\alpha = 2, \beta = 1, \gamma = 1\}$).

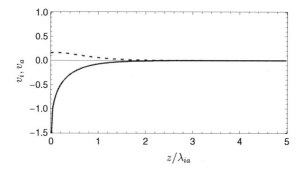

Figure 5.7.: Velocities for the unique solution. The solid line shows the ion velocity, the dashed line shows the neutral velocity ($\{\alpha = 2, \beta = 1, \gamma = 1\}$).

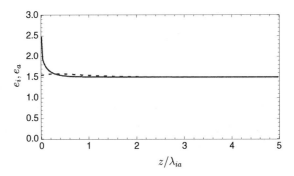

Figure 5.8.: Total kinetic energy for the unique solution. The solid line shows the ion energy, the dashed line shows the neutral energy ($\{\alpha = 2, \beta = 1, \gamma = 1\}$).

5.4.2. Example II: $\{\alpha = 0.4, \beta = 1, \gamma = 0.4\}$

This example is equivalent to example III of section 3.2. Both results (figures 5.9 and 5.10) look very similar and exhibit no problematic behavior. It can be observed that the kinetic model yields a lower neutral velocity at the cathode edge than the expanded fluid model.

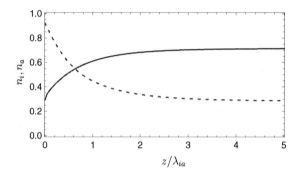

Figure 5.9.: Densities for the unique solution. The solid line shows the ion density, the dashed line shows the neutral density ($\{\alpha = 0.4, \beta = 1, \gamma = 0.4\}$).

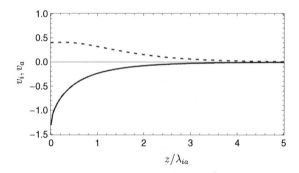

Figure 5.10.: Velocities for the unique solution. The solid line shows the ion velocity, the dashed line shows the neutral velocity ($\{\alpha = 0.4, \beta = 1, \gamma = 0.4\}$).

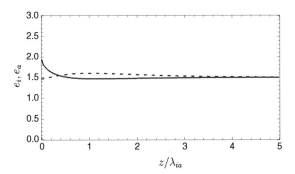

Figure 5.11.: Total kinetic energy for the unique solution. The solid line shows the ion energy, the dashed line shows the neutral energy ($\{\alpha = 0.4, \beta = 1, \gamma = 0.4\}$).

5.4.3. Example III: $\{\alpha = 0.1, \beta = 1, \gamma = 0.4\}$

The parameter combination in this example is equivalent to the combination in example IV of section 3.2. There, the expanded fluid model was not able to yield a physical solution. This problem is obviously not present anymore with the kinetic model. The neutrals are decelerated close before the cathode and thus do not reach supersonic speeds, as it can be observed in figure 5.12 and especially in figure 5.13.

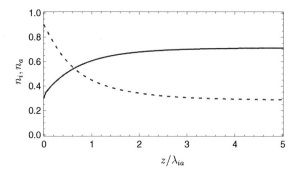

Figure 5.12.: Densities for the unique solution. The solid line shows the ion density, the dashed line shows the neutral density ($\{\alpha = 0.1, \beta = 1, \gamma = 0.4\}$).

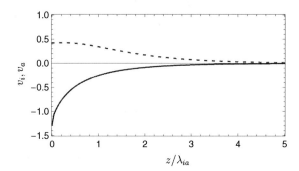

Figure 5.13.: Velocities for the unique solution. The solid line shows the ion velocity, the dashed line shows the neutral velocity ($\{\alpha = 0.1, \beta = 1, \gamma = 0.4\}$).

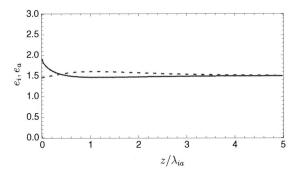

Figure 5.14.: Total kinetic energy for the unique solution. The solid line shows the ion energy, the dashed line shows the neutral energy ($\{\alpha = 0.1, \beta = 1, \gamma = 0.4\}$).

5.4.4. Example IV: $\{\alpha = 10^{-3}, \beta = 1, \gamma = 10^{-3}\}$

In this example, the parameters were chosen such that α and γ are very small and thus well in the region problematic for fluid models. The solutions from the kinetic model, however, exhibit no problematic behavior. The neutrals are decelerated just before the cathode and do not reach supersonic speeds. They also are declerated from increasing collisions with ions towards the Saha plasma, where they eventually reach zero velocity, as it can be observed in figure 5.15 and especially in figure 5.16.

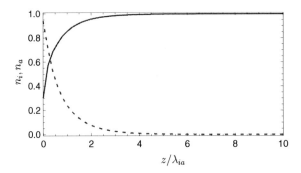

Figure 5.15.: Densities for the unique solution. The solid line shows the ion density, the dashed line shows the neutral density ($\{\alpha = 10^{-3}, \beta = 1, \gamma = 10^{-3}\}$).

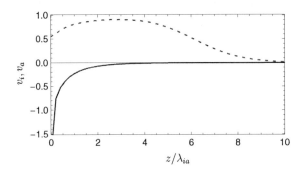

Figure 5.16.: Velocities for the unique solution. The solid line shows the ion velocity, the dashed line shows the neutral velocity ($\{\alpha = 10^{-3}, \beta = 1, \gamma = 10^{-3}\}$).

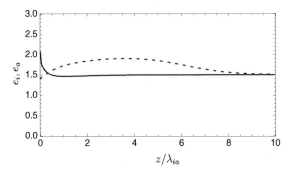

Figure 5.17.: Total kinetic energy for the unique solution. The solid line shows the ion energy, the dashed line shows the neutral energy ($\{\alpha = 10^{-3}, \beta = 1, \gamma = 10^{-3}\}$).

6. Summary and Outlook

It was found, from chapters 2 and 3, that fluid dynamic models of the near-cathode region in thermal plasmas lead to problems with respect to the existence and validity of solutions. One particular problem was the presence of supersonic neutral atoms in the solutions of both fluid models. The fluid dynamic approach itself was found to be the reason for these problems, since the assumptions necessary to justify a fluid approach are not valid near the cathode.

From this conclusion, a kinetic model was developed in chapter 4. The kinetic model was solved numerically by use of an orthonormal polynomial expansion featuring Hermite and Laguerre polynomials. Solutions for exemplary parameter combinations were calculated (cf. chapter 5.4). The results of the kinetic model and the fluid dynamic model were found to be in good agreement for combinations that were unproblematic in the fluid dynamic case. For combinations where the fluid dynamic modeling failed to yield a solution, the kinetic model was found to yield reasonable solutions; no supersonic neutrals were observed.

In addition to this obvious advantage of the kinetic model over the fluid dynamic model, the kinetic model is able to calculate additional physical quantities of interest such as the energy flux. Further advantages include a thermal coupling between plasma and cathode, as well as a larger number of parameters to control. (Which means better adaptability to given physical conditions.) Overall, the kinetic model represents a big improvement over the fluid dynamic model.

It is now necessary to further investigate the kinetic model with respect to physical validity. Parameter studies need to be carried out to determine whether the borderline between problematic and unproblematic parameter combinations has been removed or simply shifted. An initial step towards answering this question has been taken by example IV (cf. figures 5.15 and 5.16). Although example IV represents only a single parameter combination, the fact that no problems occur for such small values of α and γ is an indicator that the problems have been truely removed.

After the existence of solutions for physically reasonable parameter combinations has been ascertained, the interplay of the available parameters can be assessed further. To address this issue it will be worth the effort to implement a more advanced discretization and relaxation algorithm. Possible improvements of the solution method include a non-uniform grid (maybe even adaptive) and a non-oscillatory solution method for the discretized coefficients. The foundations to apply such improvements are already included in the solution approach described in chapter 5 due to the rather general form. Also, simply by implementing the algorithm in a programming language that is at a lower level than Mathematica (such as C++), one can expect a speed-up factor of at least 10. Such improvements were not performed here, because this work focuses on the analytical deduction of the model and the corresponding non-discretized coefficient equations.

Furthermore, the model can be expanded to allow for multiple ionization or different species. This is, however, a rather complex task. To implement it, one needs to introduce one additional distribution function for each kind of particle added. For example, implementing double ionization would require one additional distribution function; implementing a new species would require two additional distribution functions (neutrals and ions) and implementing both a new species and double ionization would require four additional distribution functions. In addition to this quickly rising number of equations, the complexity of the equations also increases. One needs to adjust the equation for the electric field (equation (4.26)) to allow for the effect of additional ions (possibly with $Z > 1$ for multiple ionization). Even more important, additional terms for the elastic collisions need to be calculated, according to the matrices in section 5.1.2. Altogether, these tasks – although mostly straightforward – heavily increase the complexity of the model and the time it takes to solve it numerically. An improved numerical implementation as mentioned in the previous paragraph is therefore mandatory before expanding the model.

A. Appendix

A.1. Saha Plasma

A Saha plasma is a plasma where ionization and recombination are in equilibrium. When the energy levels are Boltzmann distributed, this leads to the Saha equation. For one kind of ion and a given temperature T, it reads

$$S(T) = \frac{N_i N_e}{N_a} = \frac{Z_i}{Z_a} \frac{2(2\pi m_e T)^{3/2}}{h^3} e^{-E_i/T}, \tag{A.1}$$

where N represents the respective number of particles per volume element h^3, Z is the respective partition function, and E_i the ionization energy.

A.2. Behavior of the Ion Distribution Function near the Bohm Point $(z = 0)$

The Bohm point is defined as the beginning of the space charge sheath in front of a wall. The assumption of quasi-neutrality breaks down at this point. In the geometry employed in this work, this point is at the left edge of the modelled region $(z = 0)$. A completely absorbing wall is assumed, which means that all ions enter the space charge sheath with a certain distribution and eventually reach the cathode where they are neutralized and never return. For fluid dynamic models, this condition can be realized quite simply by a boundary value for the ion velocity, derived directly from the common Bohm criterion, as demonstrated in chapters 2 and 3.

For a kinetic model, the common Bohm criterion is not sufficient anymore, and a more general sheath criterion found by Riemann [32] has to be applied. To this aim, the behavior of the coefficients close to the sheath edge has to be investigated. The slope of the ion velocity decreases steeply and becomes singular at the sheath edge itself. A reparametrization of z allows to continue v_i through the singularity, and it shows that

115

A. Appendix

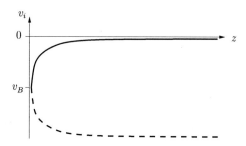

Figure A.1.: Behavior of the ion velocity near the Bohm point (at $z = 0$). The behavior is quadratic and exhibits a singularity in the slope at $z = 0$. Only the upper, solid part of the curve is physical.

v_i starts turning around, as shown in figure A.1. This second branch is of course not physical but helpful for the further mathematical treatment.

It shows that near the singularity, the dependence of the ion velocity on the variable z is quadratic. By introducing a new variable $\zeta = \sqrt{z - z_{\text{Bohm}}}$, the dependence becomes linear and can be treated more easily. Since the sheath is supposed to start at $z = 0$, the constant z_{Bohm} is equal to zero and $\zeta = \sqrt{z}$. This transformation is applied to equations (4.50) and (4.51). The derivative with respect to z is transformed according to

$$\frac{\partial f}{\partial z} = \frac{\mathrm{d}\zeta}{\mathrm{d}z} \frac{\partial f}{\partial \zeta} = \frac{1}{2\zeta} \frac{\partial f}{\partial \zeta},$$

and one obtains:

$$v_{\parallel} \frac{1}{2\zeta} \frac{\partial f_i}{\partial \zeta} - \frac{\beta}{n_i} \frac{1}{2\zeta} \frac{\mathrm{d}n_i}{\mathrm{d}\zeta} \frac{\partial f_i}{\partial v_{\parallel}} = <f_i>_c \qquad \text{(A.2)}$$

$$v_{\parallel} \frac{1}{2\zeta} \frac{\partial f_a}{\partial \zeta} = <f_a>_c \qquad \text{(A.3)}$$

Since the details of the collision terms are not of importance at this point, they have been abbreviated by $<f_i>_c$ and $<f_a>_c$. As a next step, the functions f_i, f_a, and n_i are expanded into exponential series in powers of ζ. Derivatives with respect to ζ can then be carried out directly. The expansion and the resulting derivatives read as follows:

$$f_{i/a} = f_{i/a}^{(0)} + \zeta f_{i/a}^{(1)} + \zeta^2 f_{i/a}^{(2)} + \mathcal{O}(\zeta^3) \tag{A.4}$$

$$n_i = n_i^{(0)} + \zeta n_i^{(1)} + \zeta^2 n_i^{(2)} + \mathcal{O}(\zeta^3) \tag{A.5}$$

$$\frac{\partial f_{i/a}}{\partial \zeta} = f_{i/a}^{(1)} + 2\zeta f_{i/a}^{(2)} + \mathcal{O}(\zeta^2) \tag{A.6}$$

$$\frac{\mathrm{d}n_i}{\mathrm{d}\zeta} = n_i^{(1)} + 2\zeta n_i^{(2)} + \mathcal{O}(\zeta^2) \tag{A.7}$$

With this expansion one obtains, after multiplying equation (A.2) with $2\zeta n_i$ and equation (A.3) with 2ζ:

$$v_\parallel \left(f_i^{(1)} + 2\zeta f_i^{(2)} \right) \left(n_i^{(0)} + \zeta n_i^{(1)} + \zeta^2 n_i^{(2)} \right) - \beta \left(n_i^{(1)} + 2\zeta n_i^{(2)} \right) \frac{\partial (f_i^{(0)} + \zeta f_i^{(1)} + \zeta^2 f_i^{(2)})}{\partial v_\parallel}$$
$$= 2\zeta \left(n_i^{(0)} + \zeta n_i^{(1)} + \zeta^2 n_i^{(2)} \right) < (f_i^{(0)} + \zeta f_i^{(1)} + \zeta^2 f_i^{(2)}) >_c \tag{A.8}$$

$$v_\parallel \left(f_a^{(1)} + 2\zeta f_a^{(2)} \right) = 2\zeta < (f_a^{(0)} + \zeta f_a^{(1)} + \zeta^2 f_a^{(2)}) >_c \tag{A.9}$$

All terms are regrouped according to their exponential order in ζ. From the result, one can obtain a set of equations by comparing coefficients of powers of ζ. Beginning with ζ^0, one finds:

$$v_\parallel f_i^{(1)} n_i^{(0)} - \beta \frac{\partial f_i^{(0)}}{\partial v_\parallel} n_i^{(1)} = 0 \tag{A.10}$$

$$v_\parallel f_a^{(1)} = 0 \tag{A.11}$$

The second equation is obviously fulfilled by $f_a^{(1)} = 0$, which closes the expansion for f_a. The first equation can be divided by $v_\parallel$ and then be integrated over all velocities. This integration transforms all f_i into n_i, whereas the n_i themselves remain unchanged, and one obtains a new expression for $n_i^{(0)}$:

$$n_i^{(0)} = \beta \int \frac{1}{v_\parallel} \frac{\partial f_i^{(0)}}{\partial v_\parallel} \mathrm{d}^3 v \tag{A.12}$$

This expression is equivalent to the generalized sheath criterion formulated by Riemann (equation (64) in [32]). Since the integral must not diverge, the lowest possible order of $v_\parallel$ contained in $\partial f_i^{(0)}/\partial v_\parallel$ is $v_\parallel{}^1$. This requirement has to be taken into account when

A. Appendix

expanding f_i into Hermite and Laguerre polynomials further below. The integration can be carried out by equations (A.33) and (A.34), yielding a coefficient version of the generalized sheath criterion:

$$
\begin{aligned}
n_i^{(0)} = a_{0,0}(0) &= \beta \int \frac{1}{v_\parallel} \frac{\partial f_i^{(0)}}{\partial v_\parallel} \mathrm{d}^3 v \\
&= \beta \int \frac{1}{v_\parallel} \frac{\partial}{\partial v_\parallel} \sum_{m=0}^{M} \sum_{n=0}^{N} \frac{a_{m,n}(0)}{\sqrt{2\pi}^3} \mathrm{He}_m(v_\parallel) \mathrm{La}_n(v_\perp^2) e^{-\left(v_\parallel^2 + v_\perp^2\right)/2} \mathrm{d}^3 v \\
&= \beta \int_{-\infty}^{\infty} \frac{1}{v_\parallel} \frac{\partial}{\partial v_\parallel} \sum_{m=0}^{M} \frac{a_{m,0}(0)}{\sqrt{2\pi}} \mathrm{He}_m(v_\parallel) e^{-v_\parallel^2/2} \mathrm{d}v_\parallel \\
&= \beta \left(-\sqrt{2\pi} a_{0,0}(0) + \sum_{m=0}^{M} \sum_{k=0}^{[(m-1)/2]} \frac{a_{m,0}(0)\sqrt{m!}(m-2k)}{(-2)^k k!(m-2k-1)(m-2k)!!} \right)
\end{aligned}
$$

(A.13)

Without the integration, equation (A.10) yields an expression for $f_i^{(1),e}$:

$$
f_i^{(1)} = \frac{\beta}{v_\parallel} \frac{\partial f_i^{(0)}}{\partial v_\parallel} \frac{n_i^{(1)}}{n_i^{(0)}}
$$

(A.14)

The expression above depends on $n_i^{(1)}$ and the next order, ζ^1, needs to be considered to determine $n_i^{(1)}$:

$$
v_\parallel f_i^{(1)} n_i^{(1)} + 2 v_\parallel f_i^{(2)} n_i^{(0)} - 2\beta n_i^{(2)} \frac{\partial f_i^{(0)}}{\partial v_\parallel} - \beta n_i^{(1)} \frac{\partial f_i^{(1)}}{\partial v_\parallel} = 2 n_i^{(0)} < f_i >_c^{(0)}
$$

(A.15)

$< f_i >_c^{(0)}$ describes the collision term containing zeroth order distribution functions only. In the limit $v_\parallel \to 0$ the first three terms on the left side of equation (A.15) vanish. Equation (A.14) is substituted into the fourth term and one obtains:

$$
-\beta n_i^{(1)} \frac{\partial}{\partial v_\parallel} \left(\frac{\beta}{v_\parallel} \frac{\partial f_i^{(0)}}{\partial v_\parallel} \frac{n_i^{(1)}}{n_i^{(0)}} \right) \Bigg|_{v_\parallel = 0} = 2 n_i^{(0)} < f_i >_c^{(0)} \Big|_{v_\parallel = 0}
$$

(A.16)

$$
\Rightarrow \qquad n_i^{(1)} = \pm n_i^{(0)} \sqrt{\frac{-2 < f_i >_c^{(0)} \big|_{v_\parallel = 0}}{\frac{\partial}{\partial v_\parallel} \left(\frac{\beta^2}{v_\parallel} \frac{\partial f_i^{(0)}}{\partial v_\parallel} \right) \Big|_{v_\parallel = 0}}}
$$

(A.17)

The two signs in front of the square root represent the two branches shown in figure A.1. Only the positive branch is physical and the solution with the negative sign should be dropped. Assuming that $f_i^{(0)}$ is analytical around $v_\parallel = 0$, it can be described as proportional to $v_\parallel^m$. (Orders of $v_\parallel$ higher than m vanish faster for $v_\parallel \to 0$.) Since $n_i^{(1)}$ is finite, the denominator under the root must exist. This leads to a minimal value of m:

$$\lim_{v_\parallel \to 0} \frac{\partial}{\partial v_\parallel} \left(\frac{\beta^2}{v_\parallel} \frac{\partial f_i^{(0)}}{\partial v_\parallel} \right) = \text{const} \neq 0 \qquad (A.18)$$

$$\Rightarrow \qquad m \geq 3 \qquad (A.19)$$

This means that the minimal upper limit for the Hermite expansion is $M = 3$, otherwise the expansion is not able to render the Bohm criterion correctly. In principle, it would be possible to formulate a set of boundary condition for the coefficients $a_{m,n}$ from equations (A.14) and (A.17). However, the necessary calculations due to the collision term under the root usually prevent this approach from being feasible. A similar discussion of the distribution function near the Bohm point can be found in [29].

A.3. Hermite Polynomials

The Hermite polynomials are commonly defined by (see, for example, [43] or [44])

$$H_m(v) = (-1)^m e^{v^2} \frac{d^m}{dv^m} e^{-v^2} = m! \sum_{k=0}^{[m/2]} \frac{(-1)^k 2^{m-2k}}{k!(m-2k)!} v^{m-2k}, \qquad (A.20)$$

where $[m/2]$ denotes the largest integer $\leq m/2$. When n is even (odd), $H_m(v)$ is an even (odd) function. The polynomials are orthogonal over the interval $(-\infty; \infty)$ with the kernel e^{-v^2}:

$$\int_{-\infty}^{\infty} H_m(v) H_{\tilde{m}}(v) e^{-v^2} dv = 2^m m! \sqrt{\pi} \delta_{m,\tilde{m}}$$

In this work, a modified version of the Hermite polynomials is used (see, e.g., [43]):

$$He_m(v) = 2^{-m/2} \frac{H_m(v/\sqrt{2})}{\sqrt{m!}} = \frac{(-1)^m}{\sqrt{m!}} e^{v^2/2} \frac{d^m e^{-v^2/2}}{dv^m} = \sum_{k=0}^{[m/2]} \frac{(-1)^k \sqrt{m!}}{k! 2^k (m-2k)!} v^{m-2k} \qquad (A.21)$$

119

A. Appendix

They are orthonormal with respect to the weighting function $e^{-v^2/2}\left(\sqrt{2\pi}\right)^{-1}$:

$$\int_{-\infty}^{\infty} \text{He}_m(v)\text{He}_{\tilde{m}}(v)\frac{e^{-v^2/2}}{\sqrt{2\pi}}\text{d}v = \delta_{m,\tilde{m}} \tag{A.22}$$

A number of properties and identities has been discussed in the literature for both $\text{H}_m(v)$ and $\text{He}_m(v)$,for example in [www2] or [43] - [51]. The properties relevant for this work are listed below for convenience and because of differences in the polynomial definitions and some printing errors in the original references.

Combining the recurrence relations for the regular Hermite polynomials with the new definition leads to the recurrence relations for the modified version:

$$v\text{He}_m(v) = \sqrt{m}\text{He}_{m-1}(v) + \sqrt{m+1}\text{He}_{m+1}(v) \tag{A.23}$$

$$\frac{\text{dHe}_m(v)}{\text{d}v} = \sqrt{m}\text{He}_{m-1}(v) \tag{A.24}$$

For the original Hermite polynomials $\text{H}_m(v)$, shifts in v can be represented as follows:

$$\text{H}_m(v+a) = \sum_{k=0}^{m}\binom{m}{k}\text{H}_k(v)(2a)^{m-k} \tag{A.25}$$

This result can be used to reformulate shifts in v of the modified polynomials:

$$\text{He}_m(v+a) = \sum_{k=0}^{m}\binom{m}{k}2^{-(m-k)/2}\sqrt{\frac{k!}{m!}}(\sqrt{2}a)^{m-k} \tag{A.26}$$

Alternatively, shifts can be expressed in terms of the original Hermite polynomials:

$$\begin{aligned}\text{He}_m(v+a) &= \frac{2^{-m/2}}{\sqrt{m!}}\text{H}_m\left(\frac{v+a}{\sqrt{2}}\right)\\ &= \frac{2^{-m}}{\sqrt{m!}}\sum_{k=0}^{m}\binom{m}{k}\text{H}_k(v)\text{H}_{m-k}(a)\end{aligned} \tag{A.27}$$

Another orthonormality rule for integrals involving three Hermite polynomials reads:

$$\int_{-\infty}^{\infty} H_m(v) H_{\widetilde{m}}(v) H_{\widetilde{\widetilde{m}}}(v) e^{-v^2} dv$$

$$= \begin{cases} \dfrac{\sqrt{\pi} 2^s m! \widetilde{m}! \widetilde{\widetilde{m}}!}{(s-m)!(s-\widetilde{m})!(s-\widetilde{\widetilde{m}})!} & \text{for} \quad \left\{ \begin{array}{c} 2s = m+\widetilde{m}+\widetilde{\widetilde{m}} \quad \text{even} \\ \text{and} \\ m, \widetilde{m}, \widetilde{\widetilde{m}} \leq s \end{array} \right\} \\ \\ \qquad\qquad 0 & \text{else} \end{cases} \tag{A.28}$$

Combining equations (A.27) and (A.28) yields:

$$\int_{-\infty}^{\infty} \text{He}_m(v+a) \text{He}_{\widetilde{m}}(v+b) \text{He}_{\widetilde{\widetilde{m}}}(v+c) e^{-v^2} dv$$

$$= \frac{2^{-(m+\widetilde{m}+\widetilde{\widetilde{m}})}}{\sqrt{m! \widetilde{m}! \widetilde{\widetilde{m}}!}} \sum_{k_0=0}^{m} \sum_{k_1=0}^{\widetilde{m}} \sum_{k_2=0}^{\widetilde{\widetilde{m}}} \binom{m}{k_0} \binom{\widetilde{m}}{k_1} \binom{\widetilde{\widetilde{m}}}{k_2} H_{m-k_0}(a) H_{\widetilde{m}-k_1}(b) H_{\widetilde{\widetilde{m}}-k_2}(c)$$

$$\times \begin{cases} \dfrac{\sqrt{\pi} 2^{(k_0+k_1+k_2)/2} k_0! k_1! k_2!}{\prod\limits_{i=0}^{2} ((k_0+k_1+k_2)/2 - k_i)!} & \text{for} \quad \left\{ \begin{array}{c} k_0+k_1+k_2 \quad \text{even} \\ \text{and} \\ k_0, k_1, k_2 \leq (k_0+k_1+k_2)/2 \end{array} \right\} \\ \\ \qquad\qquad 0 & \text{else} \end{cases} \tag{A.29}$$

Although the Hermite polynomials can be integrated easily for a given m, it is helpful for some cases to have a general integration formula. Such a formula can easily be derived from definition (A.21):

$$\int \text{He}_m(v) dv = \int \sum_{k=0}^{[m/2]} \frac{(-1)^k \sqrt{m!}}{k! 2^k (m-2k)!} v^{m-2k} dv$$

$$= \sum_{k=0}^{[m/2]} \frac{(-1)^k \sqrt{m!}}{k! 2^k (m-2k+1)!} v^{m-2k+1} \tag{A.30}$$

The following integral can be expressed in a similar fashion, where erf(v) is the Gaussian

A. Appendix

error function:

$$
\int v \mathrm{He}_m(v) e^{-v^2/2} \mathrm{d}v = \int \frac{(-1)^m}{\sqrt{m!}} v \frac{\mathrm{d}^m e^{-v^2/2}}{\mathrm{d}v^m} \mathrm{d}v
$$

$$
= \int \sum_{k=0}^{[m/2]} \frac{(-1)^k \sqrt{m!}}{k! 2^k (m-2k)!} v^{m-2k} \mathrm{d}v
$$

$$
= \begin{cases} -e^{-v^2/2} & \text{if } m = 0 \\[2mm] -e^{-v^2/2} v + \sqrt{\dfrac{\pi}{2}} \mathrm{erf}\left(\dfrac{v}{\sqrt{2}}\right) & \text{if } m = 1 \\[4mm] e^{-v^2/2} \dfrac{(-1)^m}{\sqrt{m!}} \left(\displaystyle\sum_{k=0}^{m-1} h_{m-1,k} v^{k+1} - \sum_{k=0}^{m-2} h_{m-2,k} v^k \right) & \text{if } m \geq 2 \end{cases}
$$

$$(A.31)$$

$h_{m,k}$ are the coefficients belonging to v^k in the Hermite polynomial $\mathrm{He}_m(v)$, multiplied by $(-1)^m \sqrt{m!}$. These coefficients are also closely related to the coefficients of the Bessel polynomials (cf. [52]):

$$
h_{m,k} = \begin{cases} \dfrac{(-1)^{(3m-k)/2} m!}{k! ((m-k)/2)! \, 2^{(m-k)/2}} & \text{if } m+k = \text{even}, \ m \geq k \\[4mm] 0 & \text{else} \end{cases}
$$

$$(A.32)$$

The next integral can only be calculated for odd coefficients m, it diverges for even m:

$$
\int_{-\infty}^{\infty} \frac{1}{v} \mathrm{He}_m(v) e^{-v^2/2} \mathrm{d}v = \int_{-\infty}^{\infty} \frac{1}{v} \sum_{k=0}^{[m/2]} \frac{(-1)^k \sqrt{m!}}{k! 2^k (m-2k)!} v^{m-2k} e^{-v^2/2} \mathrm{d}v
$$

$$
= \sum_{k=0}^{[m/2]} \frac{(-1)^k \sqrt{m!}}{k! 2^k (m-2k)!} \int_{-\infty}^{\infty} v^{m-2k-1} e^{-v^2/2} \mathrm{d}v
$$

$$
= \sum_{k=0}^{[m/2]} \frac{(-1)^k \sqrt{m!}}{k! 2^k (m-2k)!} \left(-\Gamma\left(\frac{m}{2} - k\right) 2^{m/2-k-1} (-1)^{m-2k} \right)
$$

$$
= \sum_{k=0}^{[m/2]} \frac{\sqrt{2\pi m!}}{(-2)^k k! (m-2k)!} (m-2k-2)!!
$$

$$
= \sum_{k=0}^{[m/2]} \frac{\sqrt{2\pi m!}}{(-2)^k k! (m-2k)} \frac{(m-2k)!!}{(m-2k)!}
$$

$$(A.33)$$

Similarly, the following integral can only be calculated for even coefficients m, it diverges for odd m:

$$
\begin{aligned}
\int_{-\infty}^{\infty} \frac{1}{v} \frac{\mathrm{d}}{\mathrm{d}v}\left(\mathrm{He}_m(v)e^{-v^2/2}\right)\mathrm{d}v &= \int_{-\infty}^{\infty} \frac{1}{v}\left(\sqrt{m}\mathrm{He}_{m-1}(v) - v\mathrm{He}_m(v)\right)e^{-v^2/2}\mathrm{d}v \\
&= \int_{-\infty}^{\infty} \frac{\sqrt{m}}{v}\mathrm{He}_{m-1}(v)e^{-v^2/2}\mathrm{d}v - \int_{-\infty}^{\infty}\mathrm{He}_m(v)e^{-v^2/2}\mathrm{d}v \\
&= \left(\sum_{k=0}^{[(m-1)/2]} \frac{\sqrt{m}\sqrt{2\pi(m-1)!}}{(-2)^k k!(m-2k-1)} \frac{(m-2k-1)!!}{(m-2k-1)!}\right) - \sqrt{2\pi}\delta_{m,0} \\
&= \left(\sum_{k=0}^{[(m-1)/2]} \frac{\sqrt{2\pi m!}(m-2k)}{(-2)^k k!(m-2k-1)(m-2k)!!}\right) - \sqrt{2\pi}\delta_{m,0}
\end{aligned}
$$

$$(A.34)$$

The definition of the double factorial (A.48) and its relation to the regular factorial (A.50) were used in the integrals (A.33) and (A.34).

A.4. Laguerre Polynomials

The Laguerre polynomials are commonly defined by (see, for example, [43])

$$
\begin{aligned}
\mathrm{L}_n^\alpha(v) &= \frac{e^v v^{-\alpha}}{n!}\frac{\mathrm{d}^n}{\mathrm{d}v^n}(e^{-v}v^{n+k}) = \sum_{k=0}^{n}\frac{(-1)^k}{k!}\binom{n+\alpha}{n-k}v^k, & (A.35) \\
\mathrm{L}_n(v) &= \mathrm{L}_n^0(v). & (A.36)
\end{aligned}
$$

For $\alpha \neq 0$, they are often referred to as generalized or associated Laguerre polynomials. They are orthogonal over the integral $[0,\infty)$ with the kernel $v^\alpha e^{-v}$:

$$
\int_0^\infty \mathrm{L}_n^\alpha(v)\mathrm{L}_{\tilde{n}}^\alpha(v)v^\alpha e^{-v}\mathrm{d}v = \frac{(n+\alpha)!}{n!}\delta_{n,\tilde{n}}
$$

In this work, a modified version of the Laguerre polynomials is used, defined by

$$\mathrm{La}_n^\alpha(v^2) \;=\; (-1)^n \mathrm{L}_n^\alpha(v^2/2) = \sum_{k=0}^{n} \frac{(-1)^{n+k}}{2^k k!} \binom{n+\alpha}{n-k} v^{2k}, \tag{A.37}$$

$$\mathrm{La}_n(v^2) \;=\; \mathrm{La}_n^0(v^2) = (-1)^n \mathrm{L}_n(v^2/2) = \sum_{k=0}^{n} \frac{(-1)^{n+k}}{2^k k!} \binom{n}{n-k} v^{2k}. \tag{A.38}$$

With the quadratic argument $v^2/2$, one has to adjust the weighting function in the orthogonality integral for integrations over v instead of v^2. The new kernel is easily calculated to $(-1)^{-(n+\tilde n)} v e^{-v^2/2}$, as shown below:

$$\int_0^\infty \mathrm{L}_n(u) \mathrm{L}_{\tilde n}(u) e^{-u} \mathrm{d}u$$
$$= \int_0^\infty \mathrm{L}_n(v^2/2) \mathrm{L}_{\tilde n}(v^2/2) e^{-v^2/2} \mathrm{d}(v^2/2)$$
$$= \int_0^\infty \mathrm{L}_n(v^2/2) \mathrm{L}_{\tilde n}(v^2/2) e^{-v^2/2} \frac{2v}{2} \mathrm{d}v$$
$$= \int_0^\infty \frac{\mathrm{La}_n(v^2)}{(-1)^n} \frac{\mathrm{La}_{\tilde n}(v^2)}{(-1)^{\tilde n}} e^{-v^2/2} v \mathrm{d}v \tag{A.39}$$

Since this integral is only non-zero when $n = \tilde n$, the term $(-1)^{-(n+\tilde n)}$ yields unity:

$$\int_0^\infty \mathrm{La}_n(v^2) \mathrm{La}_{\tilde n}(v^2) v e^{-v^2/2} \mathrm{d}v = \delta_{n,\tilde n} \tag{A.40}$$

The corresponding recurrence relations read:

$$v\,\mathrm{La}_n(v^2) \;=\; \frac{2}{v}\left((2n+1)\mathrm{La}_n(v^2) + (n+1)\mathrm{La}_{n+1}(v^2) + n\mathrm{La}_{n-1}(v^2) \right) \tag{A.41}$$

$$\frac{\mathrm{dLa}_n(v^2)}{\mathrm{d}v} \;=\; \frac{2n}{v}\left(\mathrm{La}_n(v^2) + \mathrm{La}_{n-1}(v^2) \right) \tag{A.42}$$

A number of properties and identities has been discussed in the literature for both $\mathrm{L}_n^k(v^2)$ and $\mathrm{La}_n^k(v^2)$, for example in [www2] or [43] - [51]. The properties relevant for this work are listed below for convenience and because of differences in the polynomial definitions and some printing errors in the original references.

Shifts in the argument of the Laguerre polynomials can be expressed as follows:

$$\text{La}_n(v^2 + a) = \sum_{k=0}^{n} \text{La}_k^{-k}(v^2)\text{La}_{n-k}^{k}(a) \tag{A.43}$$

The Laguerre polynomials are related to the Hermite polynomials:

$$\begin{aligned}
H_{2n}(v) &= (-1)^n 2^{2n} n! \text{L}_n^{-1/2}(v^2) &\text{(A.44)}\\
H_{2n+1}(v) &= (-1)^n 2^{2n+1} n! v \text{L}_n^{1/2}(v^2) &\text{(A.45)}\\
He_{2n}(v) &= \frac{2^n n!}{\sqrt{(2n)!}} \text{La}_n^{-1/2}(v^2) &\text{(A.46)}\\
He_{2n+1}(v) &= \frac{2^n n!}{\sqrt{(2n+1)!}} v \text{La}_n^{1/2}(v^2) &\text{(A.47)}
\end{aligned}$$

A.5. Varia

A.5.1. Double Factorial

The double factorial is defined for integers n as [www3]:

$$n!! = \begin{cases} 1 \cdot 3 \cdot 5 \ldots (n-2) \cdot n & \text{for } n > 0 \text{ and odd} \\ 2 \cdot 4 \cdot 6 \ldots (n-2) \cdot n & \text{for } n > 0 \text{ and even} \\ 1 & \text{for } n = -1 \text{ or } n = 0 \end{cases} \tag{A.48}$$

It can be related to the Gamma function by

$$\Gamma\left(n + \frac{1}{2}\right) = \frac{(2n-1)!!}{2^n} \sqrt{\pi} \tag{A.49}$$

and fulfills the helpful identity

$$(n-1)!! = \frac{n!}{n!!}. \tag{A.50}$$

A.5.2. Vector Decomposition into Parallel and Perpendicular Components for Certain Vector Sums

In this work, the parallel and perpendicular components of vectors with the structure $(\boldsymbol{v} - \boldsymbol{\alpha} \pm \boldsymbol{\beta})$ need to be evaluated. Assuming that $\boldsymbol{\alpha}$ and $\boldsymbol{\beta}$ are given by their lengths α and β together with the angles $\{\vartheta, \varphi\}$ and $\{\vartheta', \varphi'\}$, respectively, one obtains

$$(\boldsymbol{v} - \boldsymbol{\alpha} \pm \boldsymbol{\beta})_\parallel = v_\parallel - \alpha \cos\vartheta \pm \beta \cos\vartheta' \tag{A.51}$$

for the parallel component. To determine the (squared) perpendicular component of $(\boldsymbol{v} - \boldsymbol{\alpha} \pm \boldsymbol{\beta})$, the law of cosines needs to be applied. For completeness, the angle φ_0 for $\boldsymbol{v}$ is assumed. For practical purposes (especially when integrating over all velocities $\boldsymbol{v}$), φ_0 can usually be set to zero.

$$
\begin{aligned}
(\boldsymbol{v} - \boldsymbol{\alpha} \pm \boldsymbol{\beta})_\perp^2 &= (v_\perp \cos\varphi_0 - \alpha \sin\vartheta \cos\varphi \pm \beta \sin\vartheta' \cos\varphi')^2 \\
&\quad + (v_\perp \sin\varphi_0 - \alpha \sin\vartheta \sin\varphi \pm \beta \sin\vartheta' \sin\varphi')^2 \\
&= v_\perp^2 \cos^2\varphi_0 + \alpha^2 \sin^2\vartheta \cos^2\varphi + \beta^2 \sin^2\vartheta' \cos^2\varphi' \\
&\quad - 2v_\perp \cos\varphi_0 \alpha \sin\vartheta \cos\varphi \pm 2v_\perp \cos\varphi_0 \beta \sin\vartheta' \cos\varphi' \\
&\quad \mp 2\alpha\beta \sin\vartheta \sin\vartheta' \cos\varphi \cos\varphi' \\
&\quad + v_\perp^2 \sin^2\varphi_0 + \alpha^2 \sin^2\vartheta \sin^2\varphi + \beta^2 \sin^2\vartheta' \sin^2\varphi' \\
&\quad - 2v_\perp \sin\varphi_0 \alpha \sin\vartheta \cos\varphi \pm 2v_\perp \sin\varphi_0 \beta \sin\vartheta' \cos\varphi' \\
&\quad \mp 2\alpha\beta \sin\vartheta \sin\vartheta' \sin\varphi \sin\varphi' \\
&= v_\perp^2 + \alpha^2 \sin^2\vartheta + \beta^2 \sin^2\vartheta' - 2v_\perp \alpha \sin\vartheta \cos\varphi \cos\varphi_0 \\
&\quad \pm 2v_\perp \beta \sin\vartheta' \cos\varphi' \cos\varphi_0 - 2v_\perp \alpha \sin\vartheta \sin\varphi \sin\varphi_0 \\
&\quad \pm 2v_\perp \beta \sin\vartheta' \sin\varphi' \sin\varphi_0 \mp 2\alpha\beta \sin\vartheta \sin\vartheta' \cos(\varphi - \varphi') \tag{A.52}
\end{aligned}
$$

For the special case $\boldsymbol{\alpha} = g'e/2$, $\boldsymbol{\beta} = g'e'/2$, the result can be simplified further:

$$
\begin{aligned}
\left(\boldsymbol{v} - \frac{g'e}{2} \pm \frac{g'e'}{2} \right)_\perp^2 &= v_\perp^2 + \frac{g'^2}{4} \sin^2\vartheta + \frac{g'^2}{4} \sin^2\vartheta' - 2v_\perp \frac{g'}{2} \sin\vartheta \cos\varphi \cos\varphi_0 \\
&\quad \pm 2v_\perp \frac{g'}{2} \sin\vartheta' \cos\varphi' \cos\varphi_0 - 2v_\perp \frac{g'}{2} \sin\vartheta \sin\varphi \sin\varphi_0 \\
&\quad \pm 2v_\perp \frac{g'}{2} \sin\vartheta' \sin\varphi' \sin\varphi_0 \mp 2\frac{g'^2}{4} \sin\vartheta \sin\vartheta' \cos(\varphi - \varphi') \\
&= v_\perp^2 + \frac{g'^2}{4} \left(\sin^2\vartheta + \sin^2\vartheta' \right) \mp \frac{g'^2}{2} \sin\vartheta \sin\vartheta' \cos(\varphi - \varphi') \\
&\quad - v_\perp g' \left(\sin\vartheta \cos(\varphi - \varphi_0) \mp \sin\vartheta' \cos(\varphi' - \varphi_0) \right) \tag{A.53}
\end{aligned}
$$

The following integrals are also of interest for this work, where n is a positive integer:

$$\int_0^{2\pi} \cos^{2n+1}\varphi \, d\varphi = \int_0^{2\pi} \sin^{2n+1}\varphi \, d\varphi = 0 \tag{A.54}$$

$$\int_0^{2\pi} \cos^{2n}\varphi \, d\varphi = \int_0^{2\pi} \sin^{2n}\varphi \, d\varphi = \frac{2^{1-2n}\pi(2n)!}{(n!)^2} \tag{A.55}$$

$$\int v^n e^{-v^2/2} dv = (-2)^{(n-1)/2} \text{sgn}(v)^{n+1} \Gamma\left(\frac{n+1}{2}, \frac{v^2}{2}\right) \tag{A.56}$$

$$\int_{-\infty}^{\infty} v^n e^{-v^2/2} dv = 2^{(n-1)/2}(1 + (-1)^n)\Gamma\left(\frac{n+1}{2}\right) \tag{A.57}$$

$$\int_{-\infty}^{0} v^n e^{-v^2/2} dv = 2^{(n-1)/2}(-1)^n \Gamma\left(\frac{n+1}{2}\right) \tag{A.58}$$

$\Gamma((n+1)/2, v^2/2)$ is the incomplete gamma function. The regular gamma function, $\Gamma((n+1)/2)$, can also be written as:

$$\Gamma\left(\frac{n+1}{2}\right) = \begin{cases} \sqrt{\pi}\dfrac{(n-1)!!}{2^{n/2}} & \text{for even } n \\[2ex] \left(\dfrac{n-1}{2}\right)! & \text{for odd } n \end{cases}$$

A.5.3. Vandermonde's Identity

Vandemonde's identity is given by:

$$\sum_j \binom{m}{j}\binom{n-m}{k-j} = \binom{n}{k} \tag{A.59}$$

Bibliography

[1] R Hippler, H Kersten, M Schmidt, and K H Schoenbach (eds.), *Low Temperature Plasmas 2nd edition*, Volumes 1 and 2, Wiley-VCH, Weinheim, 2008

[2] H Hess and K-D Weltmann, *Vakuum in Forschung und Praxis* **18** (2006) 7-11 DOI: 10.1002/vipr.200600293

[3] M S Benilov and A Marotta, *J. Phys. D: Appl. Phys.* **28** (1995) 1869-1882

[4] E Fischer, *Philips J. Res.* **42** (1987) 58-85

[5] P Flesch, *Selbstkonsistente Behandlung von Elektroden und Plasma in Hochdruckgasentladungslampen*, Logos Verlag, Berlin, 2000

[6] P Flesch and M Neiger, *J. Phys. D: Appl. Phys.* **35** (2002) 1681-1694

[7] P Flesch and M Neiger, *J. Phys. D: Appl. Phys.* **36** (2003) 849-860

[8] W L Bade and J M Yos, *AVCO Report RAD-TR-62-23* (1962)

[9] W Neumann, *The Mechanism of the Thermoemitting Arc Cathode*, appeared in: R Rompe and M Steenbeck (eds.), *Ergebnisse der Plasmaphysik und der Gaselektronik*, Bd. 8, Akademie Verlag, Berlin, 1987

[10] P Tielemans and F Oostvogels, *Philips J. Res.* **38** (1983) 214-223

[11] M S Benilov and M D Cunha, *J. Phys. D: Appl. Phys.* **35** (2002) 1736-1750

[12] B Rethfeld *et al.*, *J. Phys. D: Appl. Phys.* **29** (1996) 121-128

[13] R Bötticher and W Bötticher, *J. Phys. D: Appl. Phys.* **34** (2001) 1110-1115

[14] S Lichtenberg *et al.*, *J. Phys. D: Appl. Phys.* **38** (2005) 3112-3127

[15] M S Benilov, *J. Phys. D: Appl. Phys.* **41** (2008) 144001

[16] M Redwitz *et al.*, *J. Phys. D: Appl. Phys.* **39** (2006) 2160-2179

[17] *Webster's New Encyclopedic Dictionary*, Black Dog & Leventhal Publishers Inc., New York, 1993

[18] F F Chen, *Introduction to Plasma Physics and Controlled Fusion*, Plenum Press, New York, 1984

[19] M A Lieberman and A J Lichtenberg, *Principles of Plasma Discharges and Materials Processing (Second Edition)*, John Wiley & Sons, New York, 2005

[20] J A Bittencourt, *Fundamentals of Plasma Physics*, Springer-Verlag, New York, 2004

[21] R P Brinkmann, *Kinetic Description of Plasmas*, appeared in: K H Becker, U Kogelschatz, K H Schoenbach, R. J. Barker (eds.), *Non-Equilibrium Air Plasmas at Atmospheric Pressure*, IoP Publishing, Bristol, 2005

[22] B M Smirnov and H R Reiss (eds.), *Physics of Ionized Gases*, John Wiley & Sons, New York, 2001

[23] L Boltzmann, *Vorlesungen über Gastheorie I*, J. A. Barth, Leipzig, 1896

[24] L Boltzmann, *Vorlesungen über Gastheorie II*, J. A. Barth, Leipzig, 1898

[25] M S Benilov, *J. Phys. D: Appl. Phys.* **28** (1995) 286-294

[26] M S Benilov and G V Naidis, *Phys. Rev. E* **57** 2 (1998) 2230-2241

[27] M S Benilov *et al.*, *Proc. GEC2004*, MT1.002

[28] N A Almeida *et al.*, *J. Phys. D: Appl. Phys.* **37** (2004) 3107-3116 incl. online appendices

[29] H Schmitz and K-U Riemann, *J. Phys. D: Appl. Phys.* **34** (2001) 1193-1202

[30] H Schmitz and K-U Riemann, *J. Phys. D: Appl. Phys.* **35** (2002) 1727-1735

[31] M Mitchner and C H Kruger, Jr., *Partially Ionized Gases*, John Wiley & Sons, New York, 1973

[32] K-U Riemann, *J. Phys. D: Appl. Phys.* **24** (1991) 493-518

[33] E R Harrison and W B Thompson, *Proc. Phys. Soc.* **74** (1959) 145-152

[34] N A Almeida *et al.*, *Proc 28th ICPIG* (2007) ISBN 978-80-87026-01-4 pp. 1797-1800

[35] G Kühn and M Kock, *J. Phys. D: Appl. Phys.* **39** 2401-2414

[36] G Kühn and M Kock, *Phys. Rev. E* **75** 1 (2007) 016406

[37] F H Scharf *et al.*, *Proc. LS:11* (2007) ISBN 978-0-9555445-0-7 pp. 261-262

[38] L Waldmann, *Transporterscheinungen in Gasen von mittlerem Druck*, appeared in: Flügge (ed.), *Handbuch d. Physik XII*, Springer-Verlag, Berlin, 1958

[39] G A Korn and T M Korn, *Mathematical Handbook for Scientists and Engineers*, McGraw-Hill, New York, 1961

[40] E Kampke, *Differentialgleichungen: Lösungsmethoden und Lösungen*, I, B. G. Teubner, Stuttgart, 1977

[41] O Föllinger, *Regelungstechnik*, Elitera-Verlag, Berlin, 1978

[42] J Blazek, *Computational Fluid Dynamics: Principles and Applications*, Second Edition, Elsevier, Amsterdam, 2008

[43] M Abramowitz and I A Stegun (eds.), *Handbook of Mathematical Functions*, Dover Publications, New York, 1965

[44] S Roman, *The Umbral Calculus*, Dover Publications, Mineola, 2005

[45] R Askey and S Wainger, *Am. J. of Math.* **87** No. 3 (1965) 695-708

[46] G Sansone, *Orthogonal Functions*, Interscience Publishers, New York, 1959

[47] W Magnus, F Oberhettinger, and R P Soni, *Formulas and Theorems for the Special Functions of Mathematical Physics*, Springer-Verlag, New York, 1966

[48] A Erdelyi, *J. of London Math. Soc.* **13** (1938) 154-156

[49] W N Bailey, *Quart. J. of Math.*, **10** (1939) 60-66

[50] L Carlitz, *Journal London Math.* **36** (1961) 399-402

[51] L Carlitz, *Monatsh. Math.* **66** (1962) 393-396

[52] M G Kendall, *Biometrika* **44** 1/2 (1957) 270-272

Bibliography

Note: The following resources contain URL addresses leading to sites of the world wide web. Online content is known to change dynamically and the author has no influence on the future availability or validity of these addresses and their contents.

[www1]   Market study, *Optische Technologien - Wirtschaftliche Bedeutung in Deutschland* (2007) http://www.bmbf.de/pub/marktstudie-op-tech.pdf

[www2]   Weisstein, Eric W., "Hermite Polynomial." From MathWorld–A Wolfram Web Resource. http://mathworld.wolfram.com/HermitePolynomial.html

[www3]   Weisstein, Eric W. "Double Factorial. " From MathWorld–A Wolfram Web Resource. http://mathworld.wolfram.com/DoubleFactorial.html

Frank H. Scharf

Persönliche Daten

Geburtsdatum	27. November 1977
Geburtsort	Essen
E-Mail	frank.scharf@rub.de

Schule und Studium

2004 – 2008	Promotion am Lehrstuhl für Theoretische Elektrotechnik, Ruhr-Universität Bochum
2001 – 2002	Auslandsstudium an der Purdue University, IN, USA
1998 – 2003	Studium der Elektrotechnik und Informationstechnik, Diplomstudiengang, Ruhr-Universität Bochum
1988 – 1997	Burggymnasium in Essen, Abitur
1984 – 1988	Grundschule (Schillerschule) in Essen

Beruflicher Werdegang

01/04 – 11/08	Lehrstuhl für Theoretische Elektrotechnik, Ruhr-Universität Bochum, Wissenschaftlicher Mitarbeiter
04/06 – 01/07	Technische Fachhochschule Georg-Agricola zu Bochum, Lehrbeauftragter
10/02 – 06/07	Institut für das begabte Kind, Bochum, Referent
04/02 – 08/02	Department of Medicinal Chemistry and Pharmacology, Purdue University, USA, Wissenschaftlicher Programmentwickler
10/00 – 07/01	Lehrstuhl für Theoretische Elektrotechnik, Ruhr-Universität Bochum, Studentische Hilfskraft
10/99 – 09/00	Ruhr-Universität Bochum, Tutor für Informatik